GENDER-BASED VIOLENCE AND LAYERED DISASTERS

This book investigates the widespread and persistent relationship between disasters and gender-based violence, drawing on new research with victim-survivors to show how the two forms of harm constitute 'layered disasters' in particular places, intensifying and reproducing one another.

The evidence is now overwhelming that disasters and gender-based violence are closely connected, not just in moments of crisis but in the years that follow as the social, economic and environmental impacts of disasters play out. This book addresses two key gaps in research. First, it examines what causes the relationship between disasters and gender-based violence to be so widespread and so enduring. Second, it highlights victim-survivors' own accounts of gender-based violence and disasters. It does so by presenting findings from original research on cyclones and flooding in Bangladesh and the UK and a review of global evidence on the Covid-19 pandemic. Drawing on feminist theories, it conceptualises the coincidence of gender-based violence, disasters and other aggravating factors in particular places as 'layered disasters.' Taking an intersectional approach that emphasises the connections between culture, place, patriarchy, racism, poverty, settler-colonialism, environmental degradation and climate change, the authors show the significance of gender-based violence in creating vulnerability to future disasters. Forefronting victim-survivors' experiences and understandings, the book explores the important role of trauma, and how those affected go about the process of survival and recovery. Understanding disasters as layered casts light on why tackling gender-based violence must be a key priority in disaster planning, management and recovery. The book concludes by exploring critiques of existing formal responses, which often ignore or underplay gender-based violence.

The book will be of interest to all those interested in understanding the causes and impacts of disasters, as well as scholars and researchers of gender and gender-based violence.

Nahid Rezwana is Associate Professor of Geography at the University of Dhaka in Bangladesh. Her fields of interest are hazards and disaster management, gender, social inequalities, health and climate change. Her scholarship explores how and why disasters have uneven impacts on women and how these conditions could be improved with gender-sensitive disaster management.

Rachel Pain is Professor of Human Geography at Newcastle University in the UK. Her research interests include gender-based violence, disasters and recovery, community and urban trauma, the relationships between domestic and sexual abuse and terrorism/warfare, and the theory, politics and practice of participatory/co-production approaches to research, learning and action.

Routledge Studies in Hazards, Disaster Risk and Climate Change

Series Editor: Ilan Kelman, Professor of Disasters and Health at the Institute for Risk and Disaster Reduction (IRDR) and the Institute for Global Health (IGH), University College London (UCL)

This series provides a forum for original and vibrant research. It offers contributions from each of these communities as well as innovative titles that examine the links between hazards, disasters and climate change, to bring these schools of thought closer together. This series promotes interdisciplinary scholarly work that is empirically and theoretically informed, with titles reflecting the wealth of research being undertaken in these diverse and exciting fields.

Health, Wellbeing and Community Recovery in Fukushima
Edited by Sudeepa Abeysinghe, Claire Leppold, Alison Lloyd Williams and Akihiko Ozaki

Slow Disaster
Political Ecology of Hazards and Everyday Life in the Brahmaputra Valley, Assam
Mitul Baruah

Empowerment and Social Justice in the Wake of Disasters
Occupy Sandy in Rockaway after Hurricane Sandy, USA
Sara Bondesson

Climate Change and Risk in South and Southeast Asia
Sociopolitical Perspectives
Edited by Devendraraj Madhanagopal and Salim Momtaz

Gender-Based Violence and Layered Disasters
Place, Culture and Survival
Nahid Rezwana and Rachel Pain

For more information about this series, please visit: https://www.routledge.com/Routledge-Studies-in-Hazards-Disaster-Risk-and-Climate-Change/book-series/HDC

GENDER-BASED VIOLENCE AND LAYERED DISASTERS

Place, Culture and Survival

Nahid Rezwana and Rachel Pain

Routledge
Taylor & Francis Group
LONDON AND NEW YORK

Cover image: © Ashley Cooper / Getty Images

First published 2023
by Routledge
4 Park Square, Milton Park, Abingdon, Oxon OX14 4RN

and by Routledge
605 Third Avenue, New York, NY 10158

Routledge is an imprint of the Taylor & Francis Group, an informa business

British Library Cataloguing-in-Publication Data
A catalogue record for this book is available from the British Library

ISBN: 978-0-367-54577-2 (hbk)
ISBN: 978-0-367-54578-9 (pbk)
ISBN: 978-1-003-08978-0 (ebk)

DOI: 10.4324/9781003089780

Typeset in Bembo
by SPi Technologies India Pvt Ltd (Straive)

CONTENTS

LIST OF ILLUSTRATIONS

Figures

Tables

Boxes

PREFACE

I (Nahid Rezwana) was a PhD research student at Durham University, UK, when I came across an unexpected incident. I had a neighbour who was, like me, from Bangladesh. Her husband was doing a PhD, and she was taking care of their daughter. Though officially the man was my colleague, I had a good relationship with his wife rather than with him. She had been a faculty member in a private university back home and had to leave her job to accompany her husband. She used to visit our home often. From our talks, I knew that her husband was not very helpful, and she had to compromise on every family matter. One day I was invited to her house to have afternoon tea. Her husband was taking a nap and was not in a good mood as he had a supervisory meeting in the morning. He always demanded extra care and service from his wife on those meeting days. Here, I should mention that as a PhD student, I also had supervisory meetings, and my husband barely knew that. Though living in the UK, we were following our culture and supporting men at home. However, I went to my neighbour's home with my daughter and we were having a good time. Suddenly her child fell from her baby chair, which was not buckled up properly. We were shocked, and helped the child from the floor. Her lips were slightly bleeding, though the injury was not that severe. Hearing the sounds, the husband came out of his room and suddenly started shouting at his wife and kicking things on the floor. He hit his wife's back, accusing her of causing the incident, totally ignoring my presence in his house. Seeing him in that mood, my nine-year-old daughter started crying. She was shivering with fear. She had never seen anyone behaving like that, and was shocked and traumatized. I waited for a few minutes to check the girl's injury wasn't severe, and then got out of the house immediately. I had to take extra care of my daughter for the next few days to comfort her and make her understand that not all men are like that. However, I was worried about the trauma she felt. This incident was a shock for me too. I never imagined that an educated and earning woman would be in an abusive relationship. I was surprised and embarrassed to witness this secret of their lives. She never told me anything about it, though we spent many evenings talking about different problems of family life.

This incident reminded me how much domestic violence is kept hidden within the family. It made me think about the women who are living in remote locations in Bangladesh. I knew that gender-based violence (GBV) is very high in the coastal district of Barguna, which was the study area of my PhD research. Most of the women there are uneducated, non-earning and stay at home. I became very curious to know more about their lives and how they are coping with GBV. I wanted to know if the disasters they often face had any impact on GBV.

I shared this question with my PhD supervisor Professor Rachel Pain. She liked the plan and kindly agreed to work on this topic with me. We looked for an opportunity to apply for a research grant. After waiting for a few months, we planned to start the research with our own savings. However, we received some good news; listening to our plan, Mrs. Moyes (Chairperson of the Trustees of the Christopher Moyes Memorial Foundation (CMMF), UK) advised me to apply to the CMMF. We therefore drew up a research proposal on 'Gender-based violence during cyclones', with a budget plan where Professor Rachel Pain and I planned to work voluntarily. CMMF awarded the research grant, and after completing the research, we received a further grant from them to work on 'Floods and GBV in Bangladesh and UK'. These two research studies revealed detailed pictures of women's vulnerability to GBV and disasters living in remote locations and poverty. The research findings are presented in this book.

The CMMF was formally established in 2009 to celebrate the memory of Christopher Moyes (1949–2006). He was the founder of the GoAhead transport group, Chair of Durham University Council and Deputy Lieutenant of the County of Durham. He was widely respected as a quiet, effective leader who believed in making a difference. The CMMF provides funding for projects that reflect Christopher Moyes's long-standing commitment to County Durham, Durham University, and, in particular, the University's Institute of Hazard, Risk and Resilience. This Foundation has funded several research studies on disasters such as floods and cyclones in Bangladesh, volcanic eruptions in Central America, malaria in Tanzania and earthquakes in Pakistan, mostly located in the Global South. The Foundation supported the work of Heartbeat in Seaham, County Durham from 2012–16.

ACKNOWLEDGEMENTS

We are really grateful to all the members of CMMF, and especially to Mrs. Moyes for all her inspiration and help. The research grants helped us to fulfil our dream to write this book on GBV during disasters, which is such a common disaster management issue irrespective of culture and location.

We would also like to express our gratitude to the research teams in the UK and Bangladesh, especially, Raihan, Ben and Raju without whom it would have been difficult to conduct the surveys in the UK and Bangladesh successfully on this sensitive topic. They are allies for gender research. We are grateful to all the respondents who were so generous and open in sharing their very private experiences. Without their support, it would have been impossible to conduct this important research.

1

LAYERED DISASTERS AND GENDER-BASED VIOLENCE

A global perspective

Rachel Pain and Nahid Rezwana

1.1 Gender-based violence within layered disasters

The evidence is now overwhelming that disasters and gender-based violence (GBV) are closely connected. Many disasters, in different parts of the world, have been accompanied by rises in violence such as domestic violence or domestic abuse, dowry violence, child abuse, rape and sexual assault, eve-teasing and other forms of sexual harassment, forced prostitution and sex trafficking, and forced and child marriage (see Box 1.1). It has become clear from disaster research and practitioner experience in numerous countries that the intensification of GBV and the emergence of new forms takes place not only in moments of crisis, but in the weeks, months and years that follow as the social, economic and environmental impacts of disasters play out.

The Covid-19 pandemic has brought this interrelationship into even sharper relief. Within a few weeks of the pandemic taking hold in the first months of 2020, women's organisations in many countries reported elevated rates of domestic abuse and heightened problems for survivors seeking to escape their violent homes. By April 2020, domestic abuse was described as 'the shadow pandemic' by Phumzile Mlambo-Ngcuka, Executive Director of UN Women, with lockdowns in many countries and the restriction of families within their homes having rapidly become 'a perfect storm for controlling, violent behavior behind closed doors' (UNWOMEN 2020). The significance of Covid-19 as a fast-paced and truly global disaster has helped to generate closer attention to the toxic, enduring connections between disasters and GBV.

In fact, and as this book aims to show, 'disasters' are rarely harmful events that are singular, isolated, or short-term (Knowles 2014; Rivera 2020). They are almost always experienced as multiple and layered (Rezwana and Pain 2020). In introducing the term 'layered disasters', we are building on previous ideas of 'compound

DOI: 10.4324/9781003089780-1

BOX 1.1 EXAMPLES OF GENDER-BASED VIOLENCE (GBV)

Gender-based violence (GBV): Violence directed against someone because of their gender, or violence that disproportionately affects people of a particular gender. GBV may be experienced by men, women and children, as well as transgender and non-binary people. Around the world, women are considerably more at risk than men from the forms of GBV listed below.

Domestic violence and domestic abuse: A pattern of controlling and coercive behaviour involving physical, sexual, emotional and/or psychological abuse, usually against a partner or ex-partner but sometimes other family members. 'Domestic violence' and 'domestic abuse' are sometimes used interchangeably, but the second term explicitly recognises that the violence is not only physical.

Dowry violence: Violence (commonly against wives) associated with dowry payments, often in an attempt to extort further payments, or when payments are not made or not considered sufficient.

Child abuse: Intentional harm caused to a child (usually defined as someone under the age of 18) which may be physical, sexual or emotional.

Rape and sexual assault: Rape is defined in most countries as the non-consensual sexual penetration of someone's body. Sexual assault may include rape, and a range of other unwanted sexual behaviours including non-consensual sexual touching, groping or kissing, forcing someone to touch the perpetrator or witnessing sexual activity.

Sexual harassment: Unwelcome sexual behaviour that makes someone feel intimidated or degraded. It may happen in any location and may overlap with behaviours that constitute sexual assault.

Eve-teasing: A term used primarily on the Indian sub-continent for sexual comments or advances made by men to women in public space.

Forced prostitution and sex trafficking: Forcing or coercing someone into prostitution, that is, selling sex, by another person or people. In the context of this book, forced prostitution is used to describe coercion by a family member, while sex trafficking is coercion by someone outside the family.

Forced marriage and child marriage: Forcing someone to marry without their consent. A child can never consent to marriage, and so child marriage is always considered to be forced marriage and it is also child abuse (see definition above).

Victim: Someone who has been killed or is still being affected by GBV or by a disaster.

Survivor: Someone who has been affected by GBV or a disaster, but has survived and is now beyond the incident. This term is sometimes preferred by those still in a dangerous situation.

Victim-survivor: This term is sometimes used instead, indicating that there is often no clear distinction between victimhood/survival. We use these three terms throughout this book.

disasters' as the coincidence in time and space of multiple disaster events (Liu and Huang 2014). For example, the 2017 Mocoa compound disaster in Colombia was initiated by heavy rainfall which precipitated events including landslides, debris flows and mountain torrents, together causing mass destruction (Cheng et al. 2018). Layered disasters are also multiple, but our use of this term describes the long-term, pre-existing forms of harm that already reside in place. When a major disaster strikes, these existing layers of harm (that may, at first sight, appear to be unrelated) frequently compound each other, travelling alongside and winding around the disaster, both predating and being intensified by it, in a cyclical relationship that works both ways. GBV is an important and ubiquitous example, as its presence and impacts increase during and after disasters such as cyclones and flooding; but in addition, it creates significant vulnerability to disasters in the first place. This relational cycle of layered disasters is precisely why GBV should be such a priority in disaster planning, management and recovery.

There are other sources of harm, too, that are often intimately connected to disasters in this way. Common examples include poverty, armed conflict and warfare, racism, settler-colonial violence, the mass displacement of people, dispossession of resources, environmental degradation and destruction and, of course, climate change. Some of these harms may initially be perpetrated on nature, and some on people, but often, they harm both people and nature, and become impossible to disentangle (O'Lear 2016). In this way, various disasters, harm and violence are located within complex cyclical relationships that almost always cause the most damage to the poorest and most marginalised people on our planet (George 2014; Lohokare and Davar 2000; Luft 2016; Rezwana and Pain 2020; Rivera 2020). And in turn, 'social inequalities are foundational conditions that shape both disasters and environmental risks on a global scale' (Bolin and Kurtz 2018: 182). We cannot, then, consider GBV and disasters in isolation from additional harms.

To understand the relationship between disasters and GBV, we must also understand what 'normal' or 'non-disaster' conditions are like for those who will go on to survive a 'double disaster' (Bradshaw and Fordham 2015). Often, the very idea of disaster-free normality is problematic, if we think about disasters being intertwined with these other serious harms (see Lohokare and Davar 2000). In particular, when it comes to GBV and gender relations, Luft's landmark paper on racialised disaster patriarchy is instructive, suggesting that 'gendered patterns help to produce disasters *and part of what is disastrous is patriarchy*' (Luft 2016: 8, our emphasis). We interrogate this provocation in more detail later in Section 1.3. For now, it is enough to note that there is no real distinction between what are customarily referred to as 'disasters', with their own infrastructures of research, policy and planning (see Box 1.2) and GBV, which routinely wreaks havoc on a mass scale, wrecking lives, causing injury and trauma, and having long-lasting effects on the social and economic circumstances of those affected. GBV could easily be classified as a disaster in itself, in light of this description and the definition in Box 1.2. Definitions and conceptualisations of disasters have undergone many iterations over decades of disaster studies across various academic disciplines and fields of practice (see Kelman 2020b), reflecting both the proliferation of theoretical and practitioner perspectives

BOX 1.2 HAZARDS, CRISES AND DISASTERS

Hazard: A human-made or natural phenomenon or situation with the potential for harm that presents a threat or risk, with the potential to become a disaster.

Crisis: A disruption or breakdown, a period of intense difficulty or danger. 'Crisis' is often used to describe a turning point when action is needed to avert the potential of disaster.

Disaster: A catastrophic phenomenon or situation that causes great harm and exceeds the capacity to cope of the community affected.

on disasters and concurrent multiplying debates about their causality. In outlining the conceptual framing for this book, we will go on to focus on the explanations that feminist approaches to disasters provide.

For the moment, it is enough to know that, while definitions are variable and subjective, disasters are broadly agreed to be sudden calamitous events when people, communities and societies are affected, for example, by a huge loss of life, material damage and serious disruption of daily functions (see Box 1.2). The impacts of these events are so great that they exceed the ability of the community or society to cope with its own resources. Generally, disasters entail a certain *scale* of loss of life, economic output and social infrastructure, though these are often subjective thresholds (Bradshaw and Fordham 2015). It is now widely accepted that disasters are not 'events', but rather processes that may be sparked by particular hazards and crises, but that play out over a long term (Knowles 2014). Moreover, while disasters were traditionally considered to be largely 'natural' phenomena, many commentators now see them as largely or exclusively human-produced – a result of certain political, social and economic conditions, both in the settings in which they occur and the global contexts that shape these (Hewitt 1997; O'Keefe et al. 1976). As such, disasters might be considered similar to forms of harm that are closely related to humanitarian crises, political events of war and terrorism, or forms of large-scale violence. For Kelman (2020a), disasters are informed *choices* that the most powerful groups in society make with the effect of making others vulnerable. Similarly, Mileti (1999) describes disasters as designed by societies that are well able to grasp ways to avoid them, but that lack the social or political will to do so. All disasters are framed, too, by the nature of disaster planning and response itself, as we go on to discuss later in this chapter and in Chapter 7.

In this book, our goal is to go beyond empirical associations between disasters and GBV, to investigate their relationship more closely: to examine what causes this relationship to be so widespread and so enduring. Considering the pervasive nature of GBV and its important effects on the efforts of managing disaster impacts and improving women's health and lives, the book addresses a gap in the literature by providing detailed survivor accounts on GBV and disasters. Along with empirical research located in two countries, Bangladesh and the UK, this book draws together existing research to provide an overview of the persistent, though under-researched, relation between disasters and GBV around the world. It draws upon feminist analysis to conceptualise how and why GBV increases so consistently during and after disasters.

This introductory chapter explains the rationale and conceptual framing for the rest of the book. First, in Section 1.2, we summarise the evidence that demonstrates the close relationship between disasters and GBV in many parts of the world. Next, in Section 1.3, in starting to question why this relationship is so persistent, we set out the conceptual framework for the book. In Section 1.4, we provide a short introduction to the issues that are raised for disaster planning, management and recovery by GBV and disasters, and we consider that multiple actors respond to GBV, primarily survivors themselves. Section 1.5 then outlines some of the methodological issues and challenges in research on this topic, laying the groundwork for our own studies which are reported in later chapters. And finally, in Section 1.5, we offer a summary of the chapters that follow.

1.2 Disasters and gender-based violence around the world

The impacts of disasters are far from the same for everyone, and so it is important to consider their effects across scale: on individuals, households, local communities, particular social and economic groups, and wider society. Along with physical devastation and structural damage, injuries and deaths, disasters have destructive impacts on mental health, productivity, societal structures and social relations, all of which can lead to economic instability, increased poverty, displacement and migration, and the breakdown of social protection and safety nets (Sloand et al. 2015). As a large amount of previous research has shown, the combination of disaster impacts and the failure of protective systems often aggravates gender inequalities and discrimination against women, leading to increases in GBV (Enarson et al. 2018; Gaillard et al. 2015; Masson et al. 2016; Seager 2014). GBV has therefore been described as one of the greatest challenges individuals, families and communities face during any humanitarian emergency (Inter-Agency Standing Committee 2015) and the rise of GBV during disasters and its significant impacts on individuals, families and society has become a great concern for disaster managers in recent years. Different types of GBV 'may be triggered by or exacerbated by a disaster but generally reside(s) in the everyday lived reality of women and girls' (Bradshaw and Fordham 2013: 3).

While GBV occurs in every part of the world, it varies geographically due to differences in culture, social attitudes and economic conditions. In addition, local community and family power relations may reinforce GBV, prohibiting disclosure and keeping it hidden (Sultana 2011). The World Health Organization has declared violence against women to be a preventable crisis of human health as well as a violation of women's human rights (WHO 2021). They estimate that globally, about 30% of women worldwide have been subjected to either physical and/or sexual violence in their lifetime, most of which is perpetrated by intimate partners. According to the *Global Study on Homicide 2019*, in 2017, a total of 87,000 women were intentionally killed and more than half (58%) of them (or 137 women every day across the world) were killed by a member of their own family. There are continental differences in these figures, where Africa has the highest rate of intimate partner/family-related homicide (3.1 per 100,000 female populations) and America, the second highest (1.6 per 100,000 female populations). Europe has the lowest rate of intimate partner/family-related homicide (0.7 per 100,000 female populations) (UNODC 2019).

Violence against women and girls increases at a higher rate during disasters (UNWOMEN 2014). According to the International Federation of Red Cross and Red Crescent Society (2015), globally, women and children are 14 times more likely to die or experience extreme violence in disaster situations (CCJH 2018). Women, men, girls and boys have different problems and needs during disasters, different abilities and resources to cope with their effects and different risks and vulnerabilities during disasters. Interacting with other characteristics (such as age, education, economic condition and geographical location), gender becomes a strong risk factor for unaccompanied women, girls and young boys who face different heightened risks (IFRC 2015), as three-quarters of refugees or displaced persons during post-crisis periods are female. About 1 in 5 refugee or displaced women in complex humanitarian settings have experienced sexual violence (UNICEF 2017). Women are also unsafe in their homes, as in some settings, both domestic violence and sexual violence increase after disasters (IFRC 2015; UNWOMEN 2014). During disasters, 'while men and women suffer from the same trauma, forcible displacement, injury and death, women also suffer from female-specific injustices' (Criado-Perez 2019: 296).

It is often assumed that poverty is the key exacerbating factor that explains why rates of GBV rise during and after disasters. However, while it is widely accepted that poverty has a direct relation with domestic violence, recent statistics on GBV and disasters from higher-income countries reveal that poverty is not the main cause (Bugeja et al. 2017). Post-disaster periods trigger different types of GBV and become risky for women and girls in high, middle and lower incomes countries (Hossain and McAlpine 2017; IFRC 2015; Parkinson and Zara 2013; UNICEF 2017). Presenting a literature review on disaster impacts in higher-income countries, Parkinson argues that 'gendered violence is, instead, a global concern transcending class and ethnicity' (Parkinson 2019: 2335).

For example, physical violence against women increased by 98% in the regions affected by Hurricane Katrina in the USA in 2005 (Schumacher et al. 2010). Anastario et al. (2009) studied internally displaced people (IDPs) in Mississippi after Hurricane Katrina and found that the crude rate of new cases of GBV, most commonly intimate partner violence among women, had increased from 4.6/100,000 per day to 16.3/100,000 per day by 2006 and remained elevated at 10.1/100,000 per day in 2007. In 2010, calls to the national hotline for domestic violence increased by 20% in Louisiana during the first two months after the Deepwater Horizon oil spill (US Gender & Disaster Resilience Alliance 2010 in Parkinson 2019). After analysing nine years of data from 1999 to 2007 for 67 Florida counties, Gearhart et al. (2018) found that longer-lasting exposure to natural disasters increased the expected number of assaults at the county level by approximately 78 per year.

Research elsewhere has remarkably similar findings. A study conducted in six Pacific Island countries and territories on women's lifetime experiences of violence revealed that 60–80% of the women surveyed had experienced forms of physical or sexual violence (assault, rape and sexual abuse) by partners or others, and this base level of GBV is a pervasive factor which heightens existing vulnerabilities. For example, the Tanna Women's Counselling Centre saw a 300% increase in domestic violence cases after two tropical cyclones hit Tafe Province in Vanuatu in 2011; cases of rape increased in the Solomon Islands after the Gizo tsunami in 2007; and marital rape increased in Fiji after two tropical cyclones (UNWOMEN 2014).

In the East, increases in sexual and physical violence against women were also reported in Japan during the post-disaster periods of earthquakes in 1997 and 2010 (Masson et al. 2016). New Zealand police recorded a 53% increase in domestic violence incidents over the weekend of the Canterbury earthquake in April 2010 (Phillips 2011). Domestic violence also increased in Australia after the 2009 Black Saturday bushfires, and particularly 'unwelcome findings' for Parkinson and Zara (2013: 33): 'it seemed that two decades of increased understanding of the dynamics of domestic violence by both police and the broader community was lost in this disaster's aftermath'.

According to Opondo et al. (2016), domestic violence, prostitution, child neglect and early and forced marriage increase in lower-income countries such as Uganda during droughts and prolonged dry spells. The scarcity of water in the Sahel region, Ethiopia and Papua New Guinea also leads to increased concerns about GBV, as women and girls have to walk longer distances, even at night, to fetch water (Masson et al. 2016; CARE International 2015 in Masson et al. 2016). Early forced marriage is another form of violence (see Box 1.1) which has been committed against young girls during droughts in Zimbabwe; families force their daughters' marriage to reduce family expenditure and to use dowry money which is often freed up when girls rather than women are married (Otzelberger 2014). Early marriage and leaving school early have also been observed following disasters in Bangladesh, India and Sri Lanka (UNFPA 2012 in Masson et al. 2016; Verma et al. 2013). Meanwhile, in India, human trafficking increases among disaster-hit communities, where women and girls are most often affected. Many are abducted, or exploited with the promise of false jobs or fraudulent marriages, and trafficked to other provinces or countries. Interviews in brothels in Sonagachi, Kolkata, India, revealed that many women were forced into prostitution due to regular floods in their villages (Dutta 2017).

GBV also increases during post-disaster periods in Indonesia and the Philippines. Schwefer (2018) studied 2,024 families who had been exposed to two volcanic eruptions and found that domestic violence increased by 4% among the affected families (Schwefer 2018). According to Molin (2018), although physical and sexual violence increased in the Philippines during the periods after Typhoon Bohpa in 2012 and Typhoon Haiyan in 2013 (GBV AoR 2013 in Molin 2018), awareness of

GBV was absent among many humanitarian actors, which resulted in a failure to adequately integrate GBV responses into relief programmes.

Several studies show that women are often unsafe in disaster shelters; they may try to escape from disasters, but their struggles continue when they arrive at the supposed safety of refugee camps or storm shelters. For example, women experienced sexual violence in unsegregated storm shelters in Bangladesh, and also in Louisiana's Superdome after Hurricane Katrina (Criado-Perez 2019). A study of 4,841 people after Cyclone Nargis in Myanmar in 2008 found that 31.4% felt an increased fear of sexual assault or rape, and 20.4% were worried they may be attacked at home (CCJH 2018).

These are only selected examples of the now well-established empirical relationship between disasters and GBV in different parts of the world. At the time of writing this book, the global disaster of the Covid-19 pandemic is still in full swing, with recurring waves of infection, illness and death around the world. Rises in GBV connected to the pandemic have been recorded almost everywhere. In the UK, according to Office for National Statistics (ONS 2021), the number of arrests for domestic violence-related crimes between 1 April and 30 June 2020 increased by 24% compared with the same period in the previous year. In Bangladesh, rates of GBV also increased during the pandemic period (Ehsan and Jahan 2020). According to a survey of 65,000 women and children in Bangladesh, 61% of women had faced domestic violence since Covid-19 hit the country, with 30% experiencing it for the first time during the pandemic (Dhaka Tribune 2021). According to the Dhaka Metropolitan Police (DMP), cases of violence against women were higher in March 2020 (with 56 rape cases and 132 cases of other forms of violence against women) compared to the previous month (45 and 120, respectively). However, such statistics are not wholly reliable, as only a small percentage of cases of GBV are ever reported to the authorities, and it is thought that even fewer cases are reported because of factors such as the lockdown, fear of catching the virus, and/or doubts that justice is possible during the pandemic (Mallick 2020). Chapter 5 focuses in more depth on GBV and the Covid-19 pandemic.

After a long history of protests and campaigns against gender discrimination, and despite the rise and impacts of feminism and many new policies, practices and actions on gender awareness and equality around the world, every disaster that takes place shows how far there is to go to reach the goal of gender equality in human society. GBV is pervasive in its nature and does not respect any rules, laws or geographical boundaries. This global situation of risk in 'normal' times shows that even 'non-disaster' times may be disastrous for many women and girls. And in disaster times, GBV becomes even worse, and this worsening remains for many years, particularly where there are few serious efforts to tackle it.

1.3 Patriarchy as disastrous: explaining the relationship between disasters and gender-based violence

The aim of this book is to go beyond these empirically observed connections between disasters and GBV, and to examine the relationship more closely to try to establish *why* it occurs. In this section, we outline the conceptual framework that will help to make sense of the relationship; the insights that a feminist approach brings to theorising GBV and disasters (and this relates to Section 1.5, where we identify the additional methodological implications of this approach).

A feminist approach

The authors who go beyond noting the empirical coincidence of disasters and rises in GBV (outlined in Section 1.2) and its implication for disaster management, are often feminist researchers and practitioners. In simple terms, a feminist approach aims to explain the existence and nature of gender inequality, using a framework that gives credence to people's everyday lived experiences of inequality, and identifies the structural and institutional causes. Beyond this common frame, there are multiple feminist approaches such as Black, postcolonial, queer, indigenous, Marxist, radical and liberal strands, each produced from the lived experience and activism of women and other marginalised social groups within particular geographical and social contexts.

Gender is a primary concept that, at first glance, is ubiquitous within disaster literatures and policies. Gender refers to the norms, roles and identities that are associated with being a man, a woman or someone of another gender (Box 1.3). However, the specific ways that gender is understood in feminist theory are *not* so ubiquitous within literatures and policies on disasters. In particular, patriarchy is often absent in discussions about disasters, as a structural phenomenon and institutionally reinforced source of gender inequality (see below). In addition, disaster policy tends to see gender as almost always binary, and as a descriptor for static homogenous groups of men and women, rather than diverse groupings with changing meanings, identities and roles over time (Gaillard et al. 2015). Acknowledging this diversity, in different cultural settings and places, is crucial to understanding the impacts that disasters have on different social groups. The meaning of what it is to be a man or a woman in a disaster varies culturally and geographically for men and women, and the same is true of the identities of trans and gender-non-conforming/gender-diverse people (Gaillard et al. 2015).

In recent years, but with much longer roots in Black feminism, intersectional feminism has become a popular conceptual and practical approach that recognises key differences in experiences of gender inequality and its situation in power relations between groups of women (Anzaldua 1987; Collins 1990; Crenshaw 1991). Intersectionality describes multiple forms of social and political oppression that are mutually constitutive (Hopkins 2019) – in other words, racism, sexism, ageism and class prejudice, for example, not only work together but cannot be understood in isolation (see Box 1.3). As these categories overlap, so multiple discrimination is produced. In writing on disasters, feminists have used this concept to identify who, as well as women and girls, is most impacted, and to explore differential impacts among women and men (Jacobs 2019; Luft 2016).

Box 1.3 defines some of the themes that are central to this book.

BOX 1.3 KEY THEMES IN THIS BOOK

Gender: The norms, roles and identities that are associated with being a man, a woman or someone of another gender. Gender varies between different places and cultures around the world and, it is increasingly recognised, exists beyond the binary of male/female.

Patriarchy: The system of male domination that characterises most human societies, involving a distribution of power between men and women that is generally unequal, and buttressed by social, political and economic systems which result in gender inequality.

Intersectionality: The way that multiple forms of social and political oppression can be understood as mutually constitutive. For example, racism, sexism, ageism and class or caste prejudice may work together to produce or compound the outcomes of GBV or disasters.

Place: Particular locations that have certain characteristics. These are shaped by global and local processes in different ways, to produce unique constellations of environmental and human features, lives and experiences in each place.

Culture: Sets of traditions, behaviours, beliefs, customs and norms, held and practised by particular groups or societies, and sometimes associated with or derived from living in particular places.

Trauma: The psychological impacts of harmful events such as GBV of disasters. These impacts are not restricted to individual mental health, such as the clinical condition post-traumatic stress disorder (PTSD), but also affect other aspects of the lives of victim-survivors and those around them.

Survival: Staying alive and other aspects of maintaining life for oneself or others, in response to a crisis or chronic situation that is experienced first-hand such as GBV or a disaster. Survival is often an ongoing process that has to be continually worked on, rather than a fleeting response in a short period of time.

Recovery: The process of rebuilding futures after GBV or a disaster, or during successive waves of GBV or disasters. Recovery rarely involves a return to life exactly as it was before the GBV or disaster. Recovery work may be done by victim-survivors, or by other people, organisations or states.

The concept of patriarchy, in particular, is central to a feminist analysis. Patriarchy is the system of male domination that characterises most human societies, the unequal distribution of power between men and women that is buttressed by social, political and economic systems resulting in gender inequality (Box 1.3). Placing analytical emphasis on institutions and cultures rather than individual bodies or actions, as key to patriarchy is vital; it is a system that, we go on to argue in this book, benefits many men but not all men, and while it disadvantages many women, women are differentially affected and have varying relationships to patriarchal power. Patriarchy is not a homogenous or monolithic force, but takes different forms, especially shaped by varying geographies and cultures. So while we theorise patriarchy in this book as a key force underpinning women's experiences of GBV in Bangladesh and the UK, for example, it is always shaped by (and shapes) conditions within the places in which it operates.

Most theories of patriarchy acknowledge that it never works alone, but in tandem with other forces. These result in patriarchy not only serving women and girls badly; some men and boys, especially those holding the least social and economic capital, are vulnerable to its effects, as well as gay, queer and transgender people (Dominey-Howes et al. 2014; Enarson and Pease 2016). Moreover, patriarchy intersects with other important globally ubiquitous, locally felt processes such as capitalism, racism, colonialism, environmental destruction and ecocide (Carrigan 2015; Daley 2008; Fluri and Piedalue 2017; Katz 2008; Klein 2008; Peluso and Watts 2000; Rivera 2020), to produce specific outcomes in different places. Thus a disaster in one place may look very different to a similar disaster in another place; they are likely to share some similar causal factors but different ones too (Bolin and Kurtz 2018; Dickinson 2019); they may both result in increases in GBV but the patterning of experiences, forms and effects of GBV will be specific to each location.

For these reasons, as geographers have emphasised, it is important to examine and seek to understand both disasters *and* GBV as they occur in particular places (see Box 1.3). Moreover, places – seen as locations shaped by global and local processes in different ways, producing a unique constellation of experiences of environmental and human life – are often associated with different *cultures* (see Box 1.3). Culture is understood as a complex repertoire of resources – tradition, rituals, behaviours, values, beliefs, artefacts – that people draw from to define meaning and decide how to act (Gaillard et al. 2015). Cultures are gendered in diverse ways, in turn defining gender roles and norms.

Patriarchy, gender and disasters

Feminist commentators have, for at least three decades, sought to get beyond the simple acknowledgement that GBV increases during disasters. The longstanding tendency to examine women's issues in disasters through the lens of physical and then social vulnerability has been challenged with perspectives on gender justice (Bolin and Kurtz 2018; Enarson 2009; Fordham 2011; Jacobs 2021). As Seager (2014) notes, the problem has not been a lack of attention to gender, but the bulk of research has focused on men's and women's differential vulnerability to the specific impacts of disasters, without digging much deeper:

> The observation of such differences is now so commonplace that its character is captured in what has become a ritualized one-breath phrase, ubiquitous throughout the disaster literature, 'vulnerable groups like women'.
>
> *(Seager 2014: 265)*

In running together the words in this common phrase within disaster scholarship and management, 'vulnerable groups like women', Seager offers the reality check that such 'vulnerable groups', including women, make up a majority of the population in most communities on Earth. These groups should not, therefore, be considered special cases whose needs can be addressed *after* the main business of disaster management and planning has been attended to. They should be the main business.

Seager, like other feminist commentators, identifies the importance of understanding gender relations during non-disaster times. Hence feminist literatures affect a shift to analysing the deeper causes and complexity of the underlying gendered impacts of disasters (Enarson and Morrow 1998; Fisher 2010; Hyndman 2019; Jacobs 2019, 2021; Sultana 2010). The impacts that become noticeable after disasters have continuity with 'normal' gendered patterns wherein, as a whole, women have considerably less social, economic and political power than men within society. So women are more likely than men to be in poverty, to lack financial assets, and have less access to or control over resources; to be precarious workers; to depend on state social services and public transport, and to live in public housing; to be primary carers, and to be time poor; to experience poor health, and to depend on others due to pregnancy, recent childbirth, age or illness; and to already be at risk of violence (Bradshaw and Fordham 2015; Seager 2014; Sultana 2010). All of this increases their vulnerability when it comes to disasters. And in particular, when it comes to GBV following disasters, feminists are united in pointing to the causal mechanisms underlying these persistent inequalities: 'strongly felt values and gender ideologies supporting the notion of men controlling women (and non-conforming men) are at the core of the violence' (Enarson et al. 2018: 208).

Hence, patriarchy is at the core of the analysis in this book. In particular, Luft's (2016) concept of disaster patriarchy is central to our framework. Luft's analysis of disasters centres on Hurricane Katrina which hit the Southern US in 2005. Her idea of 'racialized disaster patriarchy' intersects with Klein's (2008) analysis of disaster capitalism, which has been widely used by other critical scholars in analysing the causes of and responses to Hurricane Katrina. However, Luft centres gender relations rather than political economy in her analysis. Taking an intersectional approach (see Box 1.3), she argues that, in this context, racism and sexism worked together to produce *both* the disaster *and* the responses of formal disaster management activities and progressive social movements. The concept of disaster patriarchy therefore moves beyond the identification of gendered experiences of disasters, to examine how gendered assumptions, discrimination and oppression are part of the whole lifecourse and underlying logics of disasters:

> Racialised disaster patriarchy is deeply intersectional in its mutually-constructing, inseparable, gendered, racialized, and economic components... [it] refers to political, institutional, organisational, and cultural practices that converge before, during, and after disaster to produce intersectional gender injustice.
>
> *(Luft 2016: 3)*

Luft evidences this claim by exposing organisational and network practices and pathways, during and following the disaster, as circuits through which discourses and practices around gender frequently course. Unlike Klein's disaster capitalism, disaster patriarchy does not take advantage of shocks that wipe the slate clean; instead, it uses the opportunities presented by disaster to retrench existing inequalities of gender and race. The overall response to the disaster therefore 'returns to the most regressive elements of gender that are still embedded in social life and reanimates them', revealing that 'racialized patriarchy has been the underlying logic all along. Disaster simply unleashes, concentrates, and justifies its more prominent resurgence' (Luft 2016: 7).

For Luft, then, as previewed in the previous section of this chapter, patriarchy *itself* is disastrous. Following this inversion, patriarchy can be viewed as one of the forms of harms that we see as *interwoven* into layered disasters – harms that make disasters more frequent, prolonged, and their effects far worse for certain people in society (in this case women, and particularly racialised women). Luft's analysis underlines the importance of understanding disasters as *layered*, as whatever systems of oppression are in operation prior to disasters work to profoundly shape the path and outcomes of those disasters (Bolin and Kurtz 2018; George 2014; Jacobs 2019, 2021). And without this being a central consideration in disaster planning and management, disaster responses often fail to protect the most vulnerable people, instead making their precarious situation worse.

A disproportionate chunk of the literature on disasters analyses aspects of Hurricane Katrina, and while critical analyses of this disaster have been productive in moving analysis of disasters forward, they are inevitably also somewhat limited in their emphasis on the specific context of the US. Nonetheless, the concept of disaster patriarchy may have something to inform analysis of disasters in other parts of the world. It is important that this transference of theory moves in both directions, however, and that it is done in ways that are sensitive to and informed by the differences of place (Box 1.3).

More recently, Ensler (2021) has drawn on research in a number of countries to apply Luft's concept of disaster patriarchy to the Covid-19 pandemic. She argues that the crisis has been exploited in ways that reassert patriarchal control and dominance, erasing women's hard-earned rights, so that 'all over the world, patriarchy has taken advantage of the virus to reclaim power – on the one hand, escalating the danger and violence to women, and on the other, stepping in as their supposed controller and protector'. We discuss Covid-19, GBV and disaster patriarchy further in Chapter 5.

These are recent developments in understanding the role of patriarchy in disasters, that have been accompanied by an expansion of feminist literatures in other complementary directions. In a recent review of progress in the field of gender and disasters, Enarson et al. (2018) summarise these trends. As well as identifying the structural and institutional practices that maintain gender domination, researchers have challenged discourses around women, nature and vulnerability; begun to examine the gendered experiences of men and boys; and drawn on intersectionality as a key concept in understanding diverse gendered experiences. Poverty has emerged as a key intersecting force.

GBV, poverty and disasters

Poverty intersects with patriarchy to shape GBV and disasters and is explored in depth within the empirical studies in this book. Poverty has a direct relationship with the magnitude of the impacts of disasters (Reid 2013), as well as the increases in GBV that accompany disasters. Gender inequality and GBV are both causes and consequences of poverty. Together, they are structural forces that directly create vulnerability, especially for women and children (IFRC 2015; Opondo et al. 2016). In turn, disaster-induced economic loss in poorer societies leads to increases in violence against women and girls (Enarson 2012; Gearhart et al. 2018; Rahman 2013; Sloand et al. 2015), such as early and forced marriage (Otzelberger 2014; Verma et al. 2013), prostitution due to impoverishment (IFRC 2015), domestic violence, wife abandonment and marital rape.

These forms of violence affect women's economic status, impact their physical and mental health, and have a cascade of impacts on families and societies including low birthweight babies, higher rates of abortion and HIV, and increased demands on medical care. According to UNWOMEN (2014: 2), 'GBV does not only impact individual families and communities but also national and regional economics'. Violence hinders economic productivity, as women lose up to eight productive days after each violent incident (Opondo et al. 2016), which in turn causes more poverty at individual and societal levels, with states experiencing increased health expenditure. The cost of violence against women and girls is estimated to amount to 2% of global GDP (1.5 trillion dollars); in high-income countries, it extends to many millions, while in other countries it can be billions of dollars (Bugeja et al. 2017; UNICEF 2017). During emergencies, there is a further wide variety of direct and indirect costs of GBV, including elevated physical, mental, societal and economic costs (UNICEF 2017): 'these impacts deepen the severity of problem caused by disasters, impede recovery and threaten achievement of sustainable development' (UNWOMEN 2014: 2).

1.4 Disaster planning, management and recovery: multiple actors responding to GBV

As we have noted so far, GBV should be a key concern in the disaster responses of various actors. The bulk of the literature on these responses is concerned with the formal agents who partake in 'disaster planning' and 'disaster management', and 'disaster recovery' such as national and local governments, NGOs, and expanding private sector provision. Yet, the primary responders and actors in disaster recovery are, of course, those people and communities directly affected by disasters. For that reason, in this book, we focus *both* on policy and practice around disasters, *and* on the actions of survivors to save themselves and their loved ones and rebuild their lives, livelihoods and communities. While both disasters and GBV often create great trauma for survivors – which is compounded by a lack of support from formal actors such as the state – the women who are the primary focus of this book also engage in recovery work. This goes beyond the immediate attempt to save lives and property during disasters, to the long-term processes of survival and rebuilding. For women who experience GBV during and after disasters, with compounding effects, these actions must be undertaken to combat the impacts of both forms of harm. This perspective on survivors as disaster managers follows the call in this book to develop survivor-led understandings of GBV during disasters, underpinned by our conceptual and methodological approach (see Sections 1.3 and 1.5).

Prevailing disaster management plans and GBV

As we have argued up to this point, national-level plans and programmes to protect women from GBV during disasters are very important to reduce the spikes in violence against women that are seen when disasters hit. Among the specific factors which lead to an increase in GBV during disasters are the weakening of community and institutional protection mechanisms, limited access to justice and health services, the disruption of services and community life, the displacement and separation of families, and the loss of access to safe housing and financial assistance (Gearhart et al. 2018; IFRC 2015). Despite its importance, GBV is rarely prioritised in disaster management plans in many countries. According to an IFRC (2017) report on Ecuador, Nepal and Zimbabwe, there were no protection measures in place against the heightened risk of GBV within disaster contingency plans after major disasters such as earthquakes and drought. The national-level frameworks for GBV protection were fragile and under-resourced even in non-disaster times, and then became over-stretched in times of disaster, leaving many GBV survivors without any support (IFRC 2017). In another study conducted by IFRC (2015), nine countries (Bangladesh, Bosnia-Herzegovina, El Salvador, Haiti, Malawi, Myanmar, Namibia, Romania and Samoa) were revealed to have both national policies on disasters and national legislation on gender, and some that referred to gender in their national disaster policies; but none included arrangements for preventing GBV or even addressing GBV in their disaster plans (IFRC 2015).

Poor awareness of GBV in disasters can also be seen in high-income countries such as the USA. During Hurricane Katrina in 2005, local rape centres and domestic violence shelters had to close, which meant that survivors of GBV had no refuge during the emergency (Criado-Perez 2019). Similarly, abortion clinics were closed from the early stage of the Covid-19 pandemic in several states of the USA as they were categorised as a 'non-essential service', 'to preserve resources and protective equipment to fight the spread of the coronavirus' (Wulfhorst 2020). While affecting many women, this has particular impacts for those who suffer sexual violence during the disaster. In the UK, government action during the pandemic in response to the increased calls for support from domestic abuse survivors (an increase of 61% over the year 2020, according to Grierson 2021) was inadequate. The national government's allocation of £2 million GBP was called 'pitiful' by Fiona Dwyer, the chief executive of the NGO Solace Women's Aid (Tickle 2020). These responses during the pandemic reflect the fact that in 'normal', non-disaster times, 'women's experiences of fleeing domestic violence have been under-recognised in UK policy and practice' (Brickell and Cuomo 2020: 299). In Australia in 2012, a review of state and territory government emergency recovery policy and planning documents revealed a pervasive gender-blindness. For example, the Victorian Action Plan to Address Violence Against Women and Children includes family violence training workshops for emergency management workers, but there are still many improvements to be made to emergency management and health policies in order to improve outcomes for domestic violence survivors during post-disaster periods (Hazeleger 2013; Parkinson 2019).

It is vital to reshape formal responses to disasters so that GBV is a central consideration. Gender oppression and the urgent risks that GBV poses require intervention and prioritisation, both in disaster planning and management and in non-disaster times. At present, most interventions are inadequate to protect women, and sometimes actually worsen the impacts of disasters on them, continuing the destructive cycle of disasters, poverty and GBV. Of course, this cannot only be a requirement of policy and practice during and immediately after disasters, but must be buttressed by comprehensive prevention of GBV and support of survivors and their families. It must also be premised on an understanding of the structural causes of GBV; what are the implications for policy and practice of the concepts of layered disasters and disaster patriarchy? We discuss the implications of these arguments, and of the research presented in this book, for disaster planning, management and recovery in Chapter 7.

Trauma, survival and recovery

However, it is just as important to understand the responses of survivors of disaster-related GBV and GBV-related disasters. For that reason, this book also focuses on trauma and survival – not through an abstract lens, but from the accounts of survivors themselves. Both trauma and survival are viewed as processes that are closely connected, rather than fixed states occurring in sequence. This reflects a strengths- and empowerment-based approach that moves away from the tendency in much of the disasters literature to consider issues only in light of what institutional responses are needed to resolve them. While this is clearly an important battle, over-emphasis can have undesirable impacts on how survivors of GBV and disasters are constructed, understood and involved (or not) in policy and planning; at its worst, this perspective becomes paternalistic, centring heroic states, institutions and disaster managers as solely responsible for rescuing poor vulnerable women. As others have argued, the responses to disasters of states, NGOs and social movements have often compounded the inequalities that were imbricated in disasters' development (Klein 2008; Luft 2016). Moreover, unless women's own responses are well understood, in the context of the constraints and capacities they operate within, and particularly how these vary within particular localities and cultures, it is unlikely that policy development will be effective in resolving their problems on the ground.

For these reasons, understandings of gender and disasters have in recent years placed growing emphasis on the resilience, capacities and agency of women themselves, and to a lesser extent those of girls, boys and men (Enarson et al. 2018). Greater focus in the gender and disaster literatures on self-determination and self-organisation has also emerged. Gaillard et al. (2015: 222) describe this shift from vulnerabilities to capacities, defining the latter as 'the set of knowledge, skills and resources to which people resort in dealing with hazards and disasters'. They are clear that capacities should not be seen as the opposite of vulnerabilities, because highly vulnerable communities often have great capacities. Disaster risk reduction policies have been notoriously poor at acknowledging this, especially the influence of variable cultural gendered contexts on people's capacities (Gaillard et al. 2015).

At first glance, it may seem that this book's focus on *trauma* sits counter to this emphasis on women's capacities (Box 1.3). As the case study chapters will show, trauma has a very important role in experiences of both disasters and GBV, and in the processes of survival and recovery. A key aspect of understanding the nature and impacts of trauma forwarded here is that it is not a condition that is limited to the shock of the time of the disaster itself, or a condition that only refers to the ways that disaster shock is relived later. Trauma, as it is conceptualised here, is a social as well as a clinical phenomenon; its impacts are not restricted to individual mental health, but many aspects of survivors' lives, and beyond to families, communities and whole societies, and over time through generations (Brown 2004; Cvetkovich 2003; hooks 2003; Pain et al. 2020; Schwab 2010). This understanding is more relevant to chronic and complex forms of trauma than acute traumas such as one-off events such as car accidents. Being caught up in a disaster may lead to acute PTSD, an unpleasant set of symptoms which usually resolve over time. However, where a disaster leads to prolonged experiences of harm which are repeated and/or multiple, it is more likely to result in chronic trauma; which in turn may be understood as complex trauma, where betrayal and abuse (particularly from those expected to provide support) are present (Herman 1992; Pain 2021b). Similarly, Lohokare and Davar (2000) distinguish between 'episodic' and 'enduring' disasters, noting the differential impacts on mental health. Chronic and complex trauma have more severe mental health effects than acute trauma, are longer lasting and less amenable to conventional treatment, even in those societies where trauma treatment is readily available. Women who experience GBV after disasters are even more likely to experience serious effects of trauma than those experiencing only one of these phenomena (First et al. 2017).

Indeed, trauma is one of the mechanisms through which disasters become layered and compounded, and through which the effects of disaster continue to wind slowly on through time. A similar understanding of trauma is well established within feminist literatures on GBV (Gilfus 1999; Herman 2015), but is less often a feature of disasters literatures. Trauma, in turn, is compounded by the various effects of disasters and GBV, for example, impacts on income, livelihoods and environmental and cultural change. Survivors' experience of a lack of protection or betrayal by those who should provide support also plays an important role, for example, a family member who uses abusive behaviour at times of disaster, or a local government which fails in its duties to provide aid. However, formal responses to trauma are often very limited because 'the mainstream interventionist models in community mental health do not recognise many of these processes as [also being] "disasters"...professionals work with a narrow paradigm of "disaster" because of the philosophical basis of psychiatry in individual behaviour' (Lohokare and Davar 2000: 568).

But trauma is not simply a disenabling state. While Western models of trauma have posited it as an individualised response to calamitous events, in which the survivor becomes stuck in a cycle of repeated traumatic memories, longstanding feminist and postcolonial theories view most traumatic events as ongoing, for example, colonialism and racism (Brave Heart 2000; hooks 2003; see Pain 2021b). These theories also emphasise the agency and capacities of survivors with trauma, rather than the either/or commonly implied in mainstream literatures when it comes to being *either* traumatised *or* being capable of self and communal help. Instead, survivors are seen as *both* traumatised *and* capable. Therefore survivors of disasters, violence and other forms of harm are involved in self-help and peer support, even and especially in the most constraining and risky of circumstances (Akerkar and Fordham 2017; Pain 2021a). In this book, then, all of our case studies address the nature of trauma from disasters or from GBV, and frequently from the two combined. The survivors in our research describe the disabling and negative effects of trauma and the ways in which suffering and harm become compounded by ongoing oppression, violence and loss. However, we also hear and witness the great strength of women survivors in these situations.

At the same time as experiencing trauma, women and their communities variously work alone and together towards *survival*, which is understood as an ongoing process that has to be continually remade, rather than a fleeting response in a short period of time during a one-off incident such a cyclone (see Box 1.3). It is clear from the case studies in this book that individual women develop a myriad of coping strategies in order to protect themselves and their children, both in response to immediate threats of harm and in order to try and safeguard from future risks. Many also work with others in their communities to develop shared social reproduction activities, to rebuild communal resources after disasters or to protect those experiencing GBV. In turn, evidence from the UK suggests that these forms of caring activities contribute to mental health recovery (Akerkar and Fordham 2017).

This survival work is constant and ongoing and central to what has widely been understood as disaster *recovery* (see Box 1.3). However, if we reject framings of disasters and violence as involving only singular disruptions, then the notion of recovery as a return to previous 'normal life' becomes unworkable. Instead, new futures are rebuilt, often tentatively and precariously, and often while repeated layers of disaster are being experienced. All of these issues are discussed further in Chapter 7.

1.5 Methodological challenges and prospects

While academic and policy literatures that identify the problem of GBV during disasters have grown considerably, there is a paucity of detailed information on its nature and impacts. This urgent need is justified by the magnitude of the problem, the social and economic impacts, and the enormous human suffering that GBV during disasters causes, as we have shown, but also by the need to inform better disaster prevention and responses. In particular, as we have argued, understanding how the two forms of harm coincide in particular places and cultures is crucial. While methods to research issues of gender and disasters, more generally, have improved and diversified, including the growth of community-led research which enables greater agency and benefits for participants (Enarson et al. 2018), there is still a pressing need for the development of methodological approaches that offer new information about experience and causality (Hossain and McAlpine 2017). There is a scarcity of data, both quantitative, to establish varying rates over time, and qualitative, to help identify mechanisms in what we have suggested is a complex relationship.

This absence of detailed data on GBV after disaster is a global issue. There are a number of significant barriers to researching GBV and disasters, relating to data sources, access to survivors and ethics. Various reasons behind the lack of GBV data and the consequent invisibility of the problem is discussed in the literature (UNWOMEN 2014). These include a lack of awareness of disaster response committees that it is an issue that requires assessment and monitoring during disasters; for example, many GBV centres were closed during recent disasters in the USA (Criado-Perez 2019). In addition, there is often disruption to law-enforcement activity during emergencies, so that GBV is often not reported or recorded formally (IFRC 2015). And there may be social pressures leading to very low rates of reporting of GBV, for example, as was the case in Japan after recent earthquakes (Masson et al. 2016) and in Australia after recent bushfires (Parkinson 2019). Both low-income and higher-income countries, therefore, have insufficient data on GBV during disasters, although higher-income countries have seen more research studies than low-income countries (Bugeja et al. 2017; IFRC 2015; UNWOMEN 2014). Davies and Bennett (2016) revealed in a recent analysis published work around the time of the Zika and Ebola epidemics in 2015–2016, that less than 1% of papers on these epidemics explored the gendered impacts of the outbreaks. As Criado-Perez notes in her wide-ranging account of the ways in which men are treated as the default in data, it may be academics' fault that gender analysis is not prioritised, but also that the work of female academics tends to get less attention and fewer citations than that of males for similar reasons. As she argues, 'the gender data gap isn't just about silence. These silences, these gaps, have consequences. They impact on women's lives everyday' (Criado-Perez 2019: xii).

Even before disasters, accurate rates of GBV are often unknown. Even without the added pressures and context of emergencies, most incidents are unlikely to be reported to the authorities, even where national data collection takes place. This is a global problem but especially severe in disaster-prone countries where data and information on GBV may be limited and irregular (Rezwana 2018). Researchers have sometimes managed to circumvent this issue, through studies which seek to collect data directly. However, there are strong cultural forces that militate against survivors talking about GBV and reporting it to researchers. For example, survivors of rape and sexual assault may fear being identified for fear of social humiliation, or for reasons of self-blame; survivors of domestic abuse may fear further abuse if their partner finds out about the disclosure; children are especially powerless in these situations, and rarely report abuse to another adult unless they are supported in doing so. These challenges to research are the case everywhere but can be even more pronounced in the most disaster-prone regions.

Moreover, the methodologies used in research seeking to quantify GBV rarely produce the kinds of rich or 'thick' empirical material that will help to identify the complex mechanisms that connect disasters to GBV.

Our argument here is for research approaches and methods that centre survivor experience. Too often, the problem has only become apparent through secondary narrators such as NGOs, community leaders or healthcare workers, who do not see and may not fully understand the reach, dynamics and context of GBV. A survivor-centred approach entails developing and using a range of research techniques that are methodologically and ethically robust, in order to learn directly from survivor experiences in a way that does not cause harm or shame. This entails taking a feminist approach to research methodology (Ellsberg and Heise 2005), which complements the feminist conceptual framework that we have already outlined.

What is needed, then, are sensitive and grounded approaches that centre the accounts of survivors, in order to understand the diversity and complexity of experiences and causes of GBV in disasters. Researchers must use techniques that gain trust (and are deserving of that trust) and listen carefully to participants' accounts told with their own voices and of their own choice. Such approaches help to understand the question of *why* there is such a strong relation between GBV and disasters. They allow exploration of the additional relations with the other forms of harm that, as we have suggested in this chapter, may also be present in layered disasters. They enable us to ask people their own views and experiences of local gender cultures and other relevant contextual factors. Importantly, too, it is the best way to investigate how women are already taking action, individually and collectively, to survive GBV and disasters. All of this potentially provides rich new knowledge that is vital for more effective disaster planning and response.

Such sensitive and grounded approaches often involve qualitative methods and participatory approaches to research, and indeed, examples of these were applied in the studies reported in this book. However, in some cases, it may also mean survey work that collects quantitative as well as qualitative data. Producing quantifiable data, that allows us to assess rates and trends, is just as important a consideration in this area of research as producing qualitative data that help us to understand the mechanisms underlying those broad patterns. Finally, research in this field must focus on particular localities and particular disasters, so that we understand the context within which people are articulating their experiences. Therefore, the studies in this book draw on a range of methods, chosen and designed to maximise the chances of collecting high-quality data in the places concerned. The next three chapters present data and accounts that come directly from disaster survivors, in order to understand and centre their experiences of GBV, trauma, cyclones and flooding. The methodologies used for the studies are outlined in each chapter.

1.6 The structure of the book

The book is organised around four case studies of disasters and their relation with GBV. Each case study is intended to provide a different lens on the issue of disasters and GBV; they might be seen as transects that take one particular line through the topic. Taken together, the case studies provide us with wider insights on GBV and disasters, the roles of culture and place, and the ways in which women and others act to survive them. The case studies are on cyclones in coastal Bangladesh (Chapter 2), flooding in Rangpur (Chapter 3), flooding in Cumbria (Chapter 4) and the global Covid-19 pandemic (Chapter 5). Chapters 6 and 7 present some overall analysis from these empirical discussions, drawing together key themes, implications and recommendations.

The first three of these chapters present original findings from empirical studies, two in Bangladesh and one in the UK, which enable in-depth examinations of varied experiences of GBV, the problems and impacts it causes, as well as strategies of survival in different disaster locations. Quotations from survivors figure heavily in these chapters, as we attempt to understand their experiences. The chapter on Covid-19, in contrast, draws on multiple published and policy sources to enable a global analysis of one particular disaster, and the ways in which it has impacted GBV. Chapters 6 and 7 then embark on an analysis of the issues discovered – how we might conceptualise the relationships between disasters and GBV, and how we can understand persistent patterns of GBV in different parts of the world. They outline the limits of existing disaster management plans and strategies and make recommendations for responding to GBV connected to disasters.

Chapter 2 explores how cyclones increase GBV and the vulnerability of women to GBV in Bangladesh. Drawing on a qualitative dataset collected from intensive fieldwork in 2016 and 2017, the chapter describes the impacts of GBV on women living in the remote disaster-prone coastal region of Barguna. The victim-survivor-centred research reveals in-depth data on abuse, trafficking and domestic violence before cyclones hit (during the warning period), during cyclone events, and after cyclones (in the post-disaster period). The context of GBV in a conservative society is highlighted. The impacts of GBV on health, socio-economic position and vulnerability of women to future disasters are analysed, focusing on the strong relationships between GBV and gender relations, local culture and traditions.

Chapter 3 presents the second Bangladeshi case study, describing life in a village on the banks of River Jamuneshori in the region of Rangpur. Flooding is a regular event, and during severe floods, the whole village lies underwater. A range of types of GBV are also common in this village, and the chapter focuses on the effects on GBV when flooding occurs, drawing on empirical research that used mixed methods. Door-to-door surveys and in-depth interviews reveal the widespread presence of GBV. The analysis focuses on the complex interrelations between GBV and social, cultural and economic conditions. While both men and women are victims and survivors of GBV in this village, women experience it at much higher rates, especially during flooding. A gendered analysis of coping methods reveals the strategies used by men and women after flooding and GBV. The chapter also explores women's social and economic empowerment and how far this protects women from GBV.

Chapter 4 turns to the UK, drawing on original quantitative and qualitative research in Cumbria which examined women's experiences of the 2015 floods. The chapter first describes the context of the UK, where disasters are much less common than in Bangladesh, but GBV is a significant issue. The chapter then discusses the key findings on women's experiences in two flood-prone settlements in Cumbria. It focuses especially on the trauma and other mental health effects that followed the 2015 floods. It reports on crime experienced by women during and immediately after flooding, including some cases of GBV, as well as impacts on their finances, physical health and caring responsibilities. It shows how one trauma compounds another, in ways that amplify the layering of disasters. The research also offers some important insights on women's perceptions of the link between flooding and domestic abuse, shedding light on general understandings of GBV during disasters among these communities.

Chapter 5 examines the major global disaster of recent times – the Covid-19 pandemic. This chapter provides a global case study, drawing together existing research from around the world on the impact of Covid-19 on GBV. The pandemic has made the connection between disasters and GBV highly visible and pushed the connection between disasters and GBV more firmly onto public and policy agendas. As the chapter shows, the increase in domestic, sexual and other forms of violence against women and girls is just one of the gendered impacts of the pandemic that have fallen disproportionately on women. There have also been some unique dimensions to the devastating experiences of GBV that have accompanied the pandemic, as perpetrators have actively used the situation to control victim-survivors. Meanwhile, sources of support, whether formal or informal, have been much harder to access. The chapter assesses the social stratification of these added risks of violence, asking which women and girls are most at risk during the pandemic, and what this tells us about the intersectional nature of risk. It then analyses the rise in GBV during Covid-19 within a conceptual framework that has two parts to it. Firstly, focusing on the distinct spatial aspects of the pandemic, it explores the amplification of intimacy that produces hyper-isolation for victim-survivors. Secondly, it then considers the fit of Luft's (2016) concept of racialised disaster patriarchy to the rise of GBV during the pandemic.

Chapter 6 draws together the findings from the four case studies, to forge a deeper understanding of the relationships between GBV, disasters, place and culture.

It discusses the finding that GBV increases alongside disasters across the world, but that its forms, severity and impacts are closely related to aspects of location and place. While we discuss the local forces shaping patterns of GBV, we also highlight the importance of global structural processes. Secondly, the chapter returns to the trauma that arises from disasters and GBV, charting some of the long-term psychological and material impacts of these layered disasters as victim-survivors rebuild their lives. We highlight the effects of socio-economic status and experience of other inequalities and the availability of formal and informal support. Finally, the chapter examines the reasons, factors and contexts for the gendered nature of violence: why people become perpetrators and victim-survivors in different times and places, including during disasters. It considers why men are most likely to become perpetrators and women victim-survivors irrespective of geographical location, as well as seeks to understand the exceptions to this general pattern. It argues that GBV is historically and structurally embedded in every society.

Chapter 7 turns to disaster response and recovery. It considers what response and recovery might mean, in the context of our argument in the book so far that GBV and disasters form layered disasters because of structurally entrenched gender inequalities. First, as victim-survivors are the 'first responders' to layered disasters, the chapter considers the different ways that victim-survivors cope and rebuild from the impacts of GBV during disasters and continue their lives. Their resilience and stories of hope, within highly constrained circumstances, underline the need for improved interventions for tackling GBV within disaster management. We then explore critiques of existing formal disaster responses, management and recovery, which often ignore or underplay gendered impacts such as GBV. Following this, and partly based on what our respondents suggest would help them and their families, we offer recommendations for disaster planning and management that involve understanding local contexts, taking action to reduce vulnerability, and responding better to GBV during disasters. Due to the complex nature of GBV and its interrelationships with local cultures and traditions, however, lasting change must involve addressing the structural inequalities experienced by women, both in particular places and globally. The chapter concludes by returning to the conceptual framework around layered disasters developed in the book, highlighting the need for more radical transformation and identifying some hopeful signs in recent years.

References

Akerkar, S. & Fordham, M. (2017). "Gender, place and mental health recovery in disasters: addressing issues of equality and difference." *International Journal of Disaster Risk Reduction* 23: 218–230.

Anastario, M., Shehab, N., & Lawry, L. (2009). "Increased gender-based violence among women internally displaced in Mississippi two years post-Hurricane Katrina." *Disaster Med Public Health Preparedness* 3(1): 18–26.

Anzaldua, G. (1987). *Borderlands/La Frontera: The New Mestiza*. San Francisco, Aunt Lute Books.

Bolin, B. & Kurtz, L. C. (2018). "Race, class, ethnicity, and disaster vulnerability." In Rodriguez, H., Donner, W., & Trainor, J. E. (eds.) *Handbook of Disaster Research*. New York, Springer: 181–204.

Bradshaw, S. & Fordham, M. (2013). *Women, girls and disasters: a review for DFID*. Department for International Development (DFID), London.

Bradshaw, S. & Fordham, M. (2015). "Double disaster: disaster through a gender lens." In Shroder, J. F., Collins, A. E., Smanatha, J., Manyena, B., & Jayawickrama, J. (eds.) *Hazards, Risks and Disasters in Society*. Cambridge, Academic Press: 233–251.

Brave Heart, M. Y. H. (2000). "Wakiksuyapi: carrying the historical trauma of the Lakota." *Tulane Studies in Social Welfare* 21–22: 245–266.

Brickell, K. & Cuomo, D. (2020). "Geographies of violence: feminist geopolitical approaches." In Datta, A., Hopkins, P., Johnston, L., Olson, E., & Silva, J. M. (eds.) *Routledge Handbook of Gender and Feminist Geographies*. London and New York, Routledge: 297–307.

Brown, L. S. (2004). "Feminist paradigms of trauma treatment." *Psychotherapy: Theory, Research, Practice, Training* 41: 464–471.

Bugeja, L., Dawson, M., McIntyre, S., & Poon, J. (2017). "Domestic/family violence death reviews: an international comparison." In Dawson, M. (ed.) *Domestic Homicide and Death Review: An International Perspective*. London, UK, Palgrave Macmillan: 3–26.

CARE International (2015). *Initial Rapid Gender Assessment Report: Papua New Guinea 2015 El Nino*. Select Communities of Eastern Highlands, Morobe and Chimbu.

Carrigan, A. (2015). "Towards a postcolonial disaster studies." In DeLoughrey, E., Didur, J., & Carrigan, A. (eds.) *Global Ecologies and the Environmental Humanities: Postcolonial Approaches*. London, Routledge: 117.

CCJH (2018). Gender-based Violence in Disasters Information Sheet, Centre for Criminal Justice and Human Rights, School of Law, University College Cork, Ireland. www.ucc.ie/en/ccjhr/.

Cheng, D., Cui, Y., Su, F., Jia, Y., & Choi, C. E. (2018). "The characteristics of the Mocoa compound disaster event, Colombia." *Landslides* 15: 1223–1232.

Collins, P. H. (1990). *Black Feminist Thought: Knowledge, Consciousness and the Politics of Empowerment*. New York, Routledge.

Crenshaw, K. (1991). "Mapping the margins: intersectionality, identity politics, and violence against women of color." *Stanford Law Review* 43(6): 1241–1299.

Criado-Perez, C. (2019). *Invisible Women: Exposing Data Bias in a World Designed for Men*. London, Penguin, Random House.

Cvetkovich, A. (2003). *An Archive of Feelings: Trauma, Sexuality, and Lesbian Public Cultures*. Durham, Duke University Press.

Daley, P. (2008). *Gender and Genocide in Burundi: The Search for Spaces of Peace in the Great Lakes Region*. Oxford, James Currey.

Davies, S. & Bennett, B. (2016). *Zika and Ebola had a much worse effect on women: we need more research to address this in future*. Australia, The Conservation Media Group Ltd.

Dhaka Tribune (2021). "Domestic violence: 30% became victims during pandemic." https://www.dhakatribune.com/bangladesh/2021/03/31/30-of-domestic-violence-survivors-faced-violence-for-the-first-time-during-pandemic, accessed on 25/11/2021.

Dickinson, S. (2019). "Changing places: geographies of post-disaster landscapes." *Geography* 104(3): 116–124.

Dominey-Howes, D., Gorman-Murray, A., & McKinnon, S. (2014). "Queering disasters: on the need to account for LGBTI experiences in natural disaster contexts." *Gender, Place & Culture* 21(7): 905–918.

Dutta, P. M. (2017). "Natural disaster and vulnerability to trafficking of women and girls in India." *European Scientific Journal* 7881: 54–69.

Ehsan, S. M. A. & Jahan, F. (2020). "Analyzing the impact of Covid-19 on the mothers of Bangladesh: hearing the unheard." *Research Square*. doi:10.21203/rs.3.rs-79726/v1

Ellsberg, M. C. & Heise, L. (2005). *Researching Violence Against Women: A Practical Guide for Researchers and Activists*. Washington, DC, World Health Organization, PATH.

Enarson, E. (2009). "A gendered human rights approach to rebuilding after disaster." In Awotona, A. (ed.) *Rebuilding Sustainable Communities for Children and Their Families*. Newcastle, Cambridge Scholars Publishing: 13–38.

Enarson, E, Fothergill, A. & Peek, L. (2018). "Gender and disaster: foundations and directions for research and practice." In Rodriguez, H., Donner, W., & Trainor, J. E. (eds.) *Handbook of Disaster Research*. New York, Springer: 205–223.

Enarson, E. & Morrow, B. H. (eds.) (1998). *The Gendered Terrain of Disaster: Through Women's Eyes*. Westport, USA, Greenwood Publications.

Enarson, E. P. (2012). *Women Confronting Natural Disaster: From Vulnerability to Resilience*. Boulder, US, Lynne Rienner Publishers: 245.

Enarson, R. & Pease, B. (eds.) (2016). *Men, Masculinities and Disaster*. London, Routledge.

Ensler, E. (2021). "Disaster patriarchy: how the pandemic has unleashed a war on women." *The Guardian*, 1 June 2021.

First, J. M., First, N. L., & Houston, J. B. (2017). "Intimate partner violence and disasters: a framework for empowering women experiencing violence in disaster settings." *Affilia* 32(3): 390–403.

Fisher, S. (2010). "Violence against women and natural disasters: findings from post-tsunami Sri Lanka." *Violence against Women* 16(8): 902–918.

Fluri, J. L. & Piedalue, A. (2017). "Embodying violence: critical geographies of gender, race, and culture." *Gender, Place & Culture* 24(4): 534–544.

Fordham, M. (2011). "Gender, sexuality and disaster." In Wisner, B, Gaillard, J. C., & Kelman I.. (eds.) *The Routledge Handbook of Hazards and Disaster Risk Reduction*. London, Routledge: 395–406.

Gaillard, J. C., Fordham, M., & Sanz, K. (2015). "Culture, gender and disaster: from vulnerability to capacities." In Kruger, F., Bankoff, G., Cannon, T., Orlowski, B., & Schipper, E. L. F.(eds.) *Cultures and Disasters: Underatnding Cultural Framing in Disaster Risk Reduction*.Oxon, UK, Routledge: 222–234.

GBV AoR (Gender-based Violence Area of Responsibility) (2013). *Preventing Gender-based Violence After Typhoon Yolanda; Responding to Survivors' Needs*. http://gbvaor.net/wp-content/uploads/2015/02/Preventing-Genderbased-Violence-After-Typhoon-Yolanda-Responding-to-Survivors-needs.pdf, accessed on 02/03/2017.

Gearhart, S., Perez-Patron, M., Hammond, T. A., Goldberg, D. W., Klein, A., & Horney, J. A. (2018). "The impact of natural disasters on domestic violence: an analysis of reports of simple assault in Florida (1999–2007)." *Violence and Gender* 5(2): 87–92.

George, N. (2014). "Promoting women, peace and security in the Pacific Islands: hot conflict/slow violence." *Australian Journal of International Affairs* 68(3): 314–332.

Gilfus, M. E. (1999). "The price of the ticket: a survivor-centered appraisal of trauma theory." *Violence Against Women* 5(11): 1238–1257.

Grierson, J. (2021). "Calls to domestic abuse helpline in England up by 60% over past year." *The Guardian*, www.Theguardian.com, accessed on 20/04/2021.

Hazeleger, T. (2013). "Gender and disaster recovery: strategic issues and action in Australia." *Australian Journal of Emergency Management* 28(2): 40–46.

Herman, J. L. (1992). "Complex PTSD: a syndrome in survivors of prolonged and repeated trauma." *Journal of Traumatic Stress* 5: 377–391.

Herman, J. L. (2015). *Trauma and Recovery*. New York, Basic Books.

Hewitt, K. (1997). *Regions of Risk: A Geographical Introduction to Disasters*. England, Addison Wesley Longman.

hooks, B. (2003). "Lasting trauma." In hooks, B. (ed.) *Rock My Soul: Black People and Self Esteem*. New York, Atria Books: 21–34.

Hopkins, P. (2019). "Social geography I: Intersectionality." *Progress in Human Geography* 43(5): 937–947.

Hossain, M. & McAlpine, A. (2017). *Gender Based Violence Research Methodologies in Humanitarian Settings: An Evidence Review and Recommendations*. Cardiff, Elhra.

Hyndman, J. (2019). "Unsettling feminist geopolitics: forging feminist political geographies of violence and displacement." *Gender, Place & Culture* 26(1): 3–29.

IFRC (2015). "Unseen, Unheard: Gender-based Violence in Disasters, Global Study." *Researchers: Dzieszkowska, A., Graven Minnig, S., Lee, A., Mangwende, F., Manning, L., Nicol, Ch., Stern. S., Trapman, I.* International Federation of Red Cross and Red Crescent Societies, Geneva, Switzerland. www.ifrc.org

IFRC (2017). "Effective law and policy on gender equality and protection from sexual and gender-based violence in disasters." www.ifrc.org, accessed on 18/04/2021.

Inter-Agency Standing Committee (2015). *Guidelines for Integrating Gender-based Violence Interventions in Humanitarian Action: Reducing Risk, Promoting Resilience and Aiding Recovery*. Geneva, Switzerland, Inter-Agency Standing Committe.

Jacobs, F. (2019). "Black feminism and radical planning: new directions for disaster planning research." *Planning Theory* 18(1): 24–39.

Jacobs, F. (2021). "Beyond socail vulnerability: COVID-19 as a disaster of racial capitalism." *Sociologica* 15(1): 55–65.

Katz, C. (2008). "Bad elements: Katrina and the scoured landscape of social reproduction." *Gender Place and Culture* 15(1): 15–29.

Kelman, I. (2020a). *Disaster by Choice: How Our Actions Turn Natural Hazards into Catastrophes*. Oxford, UK, Oxford University Press.

Kelman, I. (ed.) (2020b). *Disaster Lexicon*. Version 9, 8 June 2020. Downloaded from http://www.ilankelman.org/miscellany/DisasterLexicon.doc

Klein, N. (2008). *The Shock Doctrine: The Rise of Disaster Capitalism*. London, Penguin.

Knowles, S. G. (2014). "Learning from disaster? The history of technology and the future of disaster research." *Technology and Culture* 55(4): 773–784.

Liu, M. & Huang, M. C. (2014). *Compound Disasters and Compounding Processes: Implications for Disaster Risk Management*. Geneva, Switzerland, United Nations Office for Disaster Risk Reduction.

Lohokare, M. & Davar, B. V. (2000). "Women in disasters and mental health." *Indian Journal of Social Work* 61: 565–580.

Luft, R. E. (2016). "Racialized disaster patriarchy: an intersectional model for understanding disaster ten years after Hurricane Katrina." *Feminist Formations* 28(2): 1–26.

Mallick, B. (2020). "Domsetic violence rises amid corona pandemic." *The Daily Observer*. www.observerbd.com, accessed on 23/10/2020.

Masson, V. L., Lim, S., Budimir, M., & Podboj, J. S. (2016). *Disasters and violence against women and girls. Can disasters shake social norms and power relations?* London, Overseas Development Institute.

Mileti, D. (1999). *Disasters by Design: A Reassessment of Natural Hazards in the United States*. Washington, DC, Henry Joseph Press.

Molin, J. (2018). "Preventing gender-based violence post disasters: building the capacity of humanitarian actors in the Philippines to engage with men and boys to reduce the risks of perpetration of violence." Unpublished thesis, Uppsala University, Sweden.

O'Keefe, P., Westgate, K., & Wisner, B. (1976). "Taking the naturalness out of natural disasters." *Nature* 260: 566–567.

O'Lear, S. (2016). "Climate change science and slow violence: a view from political geography and STS on mobilizing technoscientific ontologies of climate change." *Political Geography* 52: 4–13.

ONS (2021). Office for National Statistics. https://www.ons.gov.uk/peoplepopulationand community/crimeandjustice, accessed on 21/01/2021

Opondo, M., Abdi, U., & Nangiro, P. (2016). "Assessing gender in resilience programming: Uganda." *BRACED Resilience Intel* 2(2): 1–16.

Otzelberger, A. (2014). "Tackling the Double Injustice of Climate Change and Gender Inequality." CARE International.

Pain, R. (2021a). "Collective trauma? Isolating and commoning gender-based violence." *Gender, Place and Culture*. doi:10.1080/0966369X.2021.1975103.

Pain, R. (2021b). "Geotrauma: violence, place and repossession." *Progress in Human Geography* 45(5): 972–989.

Pain, R., Rezwana, N., & Sahdan, L. (2020) "Trauma, gender and space." In Datta, A., Hopkins, P, Johnston, L, Olson, E, & Silva, J. M. (eds.) *Routledge Handbook of Gender and Feminist Geographies*. London, Routledge: 287–296.

Parkinson, D. (2019). "Investigating the increase in domestic violence post disaster: an Australian case study." *Journal of Interpersonal Violence* 34(11): 2333–2362.

Parkinson, D. & Zara, C. (2013). "The hidden disaster: domestic violence in the aftermath of natural disaster." *The Australian Journal of Emergency Management* 28(2): 28–35.

Peluso, N. L. & Watts, M. (2000). *Violent Environments: Social Bonds and Racial Hubris*. Ithaca, New York, Cornell University Press.

Phillips, S. (2011). *Domestic Violence on Rise in Christchurch after Earthquake Stress (suite101. com: Insightful writers. Informed readers, 1 March 2011)*. http://www.whealth.com.au/documents/environmental/justice/women-disaster-violence-litreview.pdf.

Rahman, M. S. (2013). "Climate change, disaster and gender vulnerability: a study on two divisions of Bangladesh." *American Journal of Human Ecology* 2(2): 72–82.

Reid, M. (2013). "Disasters and social inequalities." *Sociology Compass* 7(11): 984–997.

Rezwana, N. (2018). *Disasters, Gender and Access to Healthcare: Women in Coastal Bangladesh*. Oxon, UK, Routledge, Taylor & Francis Groups.

Rezwana, N. & Pain, R. (2020). "Slow violence and layered disasters: gender-based violence before, during and after cyclones." *Disasters* 45(4): 741–761.

Rivera, D. Z. (2020). "Disaster colonialism: a commentary on disasters beyond singular events to structural violence." *International Journal of Urban and Regional Research*. doi:10. 1111/1468-2427.12950

Schumacher, J. A., Coffey, S. F., Norris, F. H., Tracy, M., Clements, K., & Galea, S. (2010). "Intimate partner violence and Hurricane Katrina: predictors and associated mental health outcomes." *Violence and Victims* 25(5): 588–603.

Schwab, G. (2010). *Haunting Legacies: Violent Histories and Transgenerational Trauma*. New York, Columbia University Press.

Schwefer, M. (2018). *Sitting on a Volcano: Domestic Violence in Indonesia Following Two Volcano Eruptions* (No. 263). Working Paper No. 263.

Seager, J. (2014). Disasters are gendered: what's new? In Singh, A. & Zommers, Z. (eds.) *Reducing Disaster: Early Warning Systems for Climate Change*. Dordrecht, Springer: 265–281.

Sloand, E., Killion, C., Gary, F. A., Dennis, B., Glass, N., Hassan, M., Campbell, D. W., & Callwood, G. B. (2015). "Barriers and facilitators to engaging communities in gender-based violence prevention following a natural disaster." *Journal of Health Care for the Poor and Underserved* 26(4): 1377–1390.

Sultana, F. (2010). "Living in hazardous waterscapes: gendered vulnerabilities and experiences of floods and disasters." *Environmental Hazards* 9(1): 43–53.
Sultana, F. (2011). "Spaces of power, places of hardship: rethinking spaces and places through a gendered geography of water." In Raju, S. (ed.) *Gendered Geographies: Interrogating Space and Place in South Asia*. Oxford, Oxford University Press: 293.
Tickle, L. (2020). "Fiona Dwyer: '£2m for coronavirus domestic abuse victims? It's pitiful'." Interview, *The Guardian*, www.Theguardian.com, accessed on 20/04/2021.
US Gender & Disaster Resilience Alliance (2010). *Gulf Coast women reflect on 5th anniversary of Hurricane Katrina: challenges of slow recovery and women's marginalisation after the recent oil spill*. https://www.gdnonline.org/resources/KatrinaWomenPlusFive-GDRA-Aug31.pdf
UNFPA (United Nations Population Fund) (2012). "Marrying Too Young: End Child Marriage." New York, UNFPA.
UNICEF (2017). "The material risks of gender-based violence in emergency settings: is GBV impacting your investment?" www.gender@unicef.org, accessed on 07/03/2021.
UNODC (2019). *Global Study on Homicide 2019*. Vienna, United Nation.
UNWOMEN (2014). "Climate change, disasters and gender-based violence in the pacific." www.asiapacific.unwomen.org, accessed on 20/03/2021.
UNWOMEN (2020). "Violence against women and girls: the shadow pandemic." https://www.unwomen.org/en/news/stories/2020/4/statement-ed-phumzile-violence-against-women-during-pandemic, accessed on 05/04/22.
Verma, R., Sinha, T., & Khanna, T. (2013). "Asia child marriage initiative: summary of research in Bangladesh, India and Nepal." *International Journal Center for Research on Women* Jan: 35–46.
World Health Organization (2021). *Factsheet: Violence Against Women*. https://www.who.int/news-room/fact-sheets/detail/violence-against-women, accessed on 05/04/22.
Wulfhorst, E. (2020). "U.S. states accused of using COVID-19 to close abortion clinics." Thomson Reuters Foundation. www.reuters.com, accessed on 20/04/2021.

2

GENDER-BASED VIOLENCE BEFORE, DURING AND AFTER CYCLONES IN THE COASTAL REGION OF BANGLADESH

Nahid Rezwana and Rachel Pain

2.1 Introduction

As we outlined in Chapter 1, gender-based violence (GBV) is one of the greatest challenges that individuals, families and communities face during disasters. However, as we also explained, GBV often remains hidden: it is unrecognised and under-reported, especially in lower-income disaster-prone countries where data on prevalence, and information on the nature and impacts of GBV, are very limited and irregular. This chapter focuses on the first of our four case studies of disasters, aiming to document and understand the different forms of GBV during the recent cyclones in Bangladesh. Drawing on a qualitative dataset collected from intensive fieldwork in 2016 and 2017, it focuses on the experiences and opinions of the victim-survivors, to understand the gendered impacts of disasters and GBV on women's lives in this particularly disaster-prone region.

2.2 Background: disasters and GBV in Bangladesh

Bangladesh is a disaster-prone country. This small country with a large population of over 164 million people is the fifth most disaster-prone country in the world (The World Bank 2016). The country suffers from tropical cyclones, floods, riverbank erosion, tornadoes, thunderstorms, arsenic contamination and earthquakes (Banglapedia 2015). Due to dense population and poor economic conditions, disasters always affect large numbers of people (Loayza et al. 2009; Matin and Taher 2001). However, disasters have gendered impacts, and the numbers of female victim-survivors are always higher. For example, during the 1991 cyclone, 90% of those killed were women and children, and in Cyclone Sidr 2007, the female to male death ratio was 5:1 (Ahmed et al. 2011; Rahman 2013). When it comes to the very significant health impacts of these disasters, women

DOI: 10.4324/9781003089780-2

are more vulnerable than men in regards to both health problems and healthcare access after disasters (Rezwana 2018).

GBV is also very common; indeed, it is a major social problem in Bangladesh. According to a recent survey, 12,749 incidences of violence were perpetrated against 57,704 women and children during the month of June 2020, which equates to 22% of the sample (Manusher Jonno Foundation 2021). According to Ain o Salish Kendra, there were 975 rape incidents reported from January to September 2020 and 43 deaths caused by rape in Bangladesh in the same period (ASK 2021). Child marriage is another major problem in this country. 'Over 50 per cent of Bangladeshi women who are now in their mid-20s were married before they turned 18. Nearly 18 per cent were below 15 years of age' (UNICEF 2021); however, child marriage is not recorded as GBV in government reports, although it is a criminal act according to the law (BBS 2013a). In addition, a high number of Bangladeshi women become subject to violence in their husbands' houses after marriage. According to a recent government report on 'Violence Against Women (VAW) Survey 2015', 80.2% of women experienced domestic abuse, including physical, sexual, economic and emotional abuse and controlling behaviour. Around 41.7% of those subjected to physical or sexual violence suffered injuries (BBS 2016). Such statistics are likely to be only the tip of the iceberg, given under-reporting. As Hossain and Suman (2013: 81) note, 'in a traditional value dependent society like Bangladesh, many incidents of violence against women are not reported to the police or in the media'. The same government report also emphasises that about 72.7% of female survivors of GBV do not share their experiences with anyone, and only 2.6% had sought legal support (BBS 2016).

These endemic conditions of GBV raise questions about the extent, nature and effects of GBV after disasters in Bangladesh. But there is no detailed information about GBV during disasters in Bangladesh, with the exception of just a few individual studies (Sultana 2009, 2010; Rashid and Michaud 2000). Bangladesh does not have records of GBV after recent cyclones like Cyclone Sidr in 2007 and Cyclone Mahasen in 2013, as no surveys have been conducted in the affected areas. However, an earlier research project that investigated women's access to healthcare during cyclone disasters revealed several cases of GBV after Cyclone Sidr, which were otherwise unreported and unrecorded. It became clear that the presence of different forms of violence created highly vulnerable conditions for the women and girls living in the remote coastal region of Bangladesh. These cases were mentioned by NGO officials working in the coastal regions, whereas local inhabitants avoided questions related to violence (Rezwana 2018). This is a mark of the hidden nature of GBV within this society, which itself increases the vulnerability of women and girls to disasters, and reduces community resilience and the efficacy of state and NGO responses. These findings were what prompted the research reported in this chapter – only a qualitative study could capture the complexities of GBV and disasters in women's lives. As Bradshaw and Fordham have noted (2013: 7), 'there is a lack of quantitative data available to measure the gendered impact of disasters, and the lack of large-scale qualitative studies by which to better understand how disasters are experienced as gendered events'.

The current chapter therefore addresses this gap, investigating the nature and effects of GBV after disasters in Bangladesh.

2.3 Aims, study area and methodology

The research reported here, carried out in 2016 and 2017, aimed to document and understand the different forms of GBV that have occurred during recent cyclone events in the coastal regions of Bangladesh. The research aimed to address a significant gap in understanding, both within research and in policy and planning around disasters. The qualitative methods used were designed to capture the problems of GBV faced by women and girls in this region, as they defined them, and to assess the effects of disasters on GBV and vice versa. The goal was to gather empirical material that would help in analysing the complex interrelations of socio-cultural factors, GBV and disasters.

The study area

The research was located in Barguna, a district on the southwest coast of Bangladesh. Barguna is cyclone-prone due to its geographical location in the risk zone for tropical cyclones (National Plan for Disaster Management 2010). It has a long history of cyclones, the strongest of which were in 1935, 1965, 1970, 2007, 2013 and 2016. However, any cyclone which passes Barguna has impacts, due to the district's dense population (total population is 892,787, population density 488 per square kilometre and the female–male ratio is 100:96) and dependency on the agricultural sector (out of the total holdings 71.93% depend on agriculture) (BBS 2013b). Most of the inhabitants belong to the country's poorest socio-economic group (Health Bulletin 2013) and women are mostly housewives (73.2%), dependent on their family income. Only 4.02% of women are employed in the formal labour force, and others work in the informal sector or do not work. The average literacy rate is 57.6% among Barguna inhabitants, but lower among females. In particular, the percentage of people with a degree (i.e. above secondary school level) is very low among the female population, at 4.78% of those aged 20–24 years, compared to 12.97% of men of the same age (BBS 2013b). Barguna, as a district, performs poorly in providing basic facilities to its inhabitants. According to government records, only 1.8% of households have a tap water supply, and 28.8% have electricity (BBS 2013b). Healthcare facilities are also insufficient in Barguna, with one doctor for every 28,810 people. There is only one District Hospital which has only 100 beds, while other government health centres offer 221 beds and 23 private clinics offer only 30 beds (Health Bulletin 2013). Along with Barguna's geographical location, then, these insufficiencies in basic facilities greatly magnify the vulnerability of the inhabitants during cyclones. In addition, socio-cultural factors including patriarchy, economic conditions and disparities between men and women in receiving information and healthcare, and gendered roles and responsibilities increase women's vulnerability in particular.

Methodology

The research took an intensive, grounded qualitative research approach, to identify and explore GBV with women and the agencies which work with them. The research approach rested on the development of trust and relationship-building with the communities affected by disasters, using qualitative methods to identify the key issues and questions. The experience of the research team in this region of Bangladesh, therefore, informed the methodology.

A number of field visits were made in 2016 and 2017 to collect data from the urban and rural areas of Barguna. The first visit, just after Cyclone Roanu in 2016, was to select Barguna as a study area and to recruit local research assistants. The second visit involved meeting with the key informants, training the research assistants, undertaking the pilot surveys, and conducting semi-structured interviews with eight key informants. The key informants included five NGO officials and three local inhabitants; most worked with GBV survivors, and two of the key informants (a male and a female) were themselves survivors of child abuse, forced marriage and domestic violence. The key informants were also asked for their input into the research questions and methodology.

Several visits to the area were then made to collect data using in-depth interviews and questionnaires. In this main phase of the research, we worked with key informants in Barguna, and through these contacts and snowballing, we recruited further respondents.

In-depth interviews and questionnaires were conducted with 29 female GBV survivors and 8 male respondents. The female respondents were all over ten years of age (as cases of child marriage start from age ten). Most of the female respondents were of very poor socio-economic status (as compared with an extreme poverty line equivalent to USD 1.9 PPP; World Bank 2020), and nearly half were housewives dependent on family income for economic support. The rest worked as day labourers, housemaids, students and NGO junior staff. Most of the male respondents were businessmen, NGO officials or teachers. Two were known to be GBV perpetrators, known in the locality for violence at home, and one was a survivor of child abuse. All these respondents shared not only their own experiences, but also examples of GBV that they knew of or had heard about. Conducting in-depth interviews was challenging due to its sensitive nature and cultural context, and so a strict ethical protocol was developed and adhered to. Bad weather and the poor transport system in and out of Barguna meant that several visits were necessary. However, the research team succeeded in gaining trust within the community, and the GBV survivors were eventually keen to share their stories.

This main phase of fieldwork at different sites in Barguna was supported with interviews with key personnel from national, municipal and local government, NGOs, educational institutes and other service providers working in the region. These generated a second dataset on GBV during disasters that explored existing responses and engaged professionals with findings and recommendations (discussed in Chapter 7) from the community stage of the research.

2.4 GBV in Barguna during non-disaster periods

Before going on to examine GBV during disasters, a description of GBV in Barguna during the rest of the year is helpful; this is the context in which disasters occur. These findings are based on the experiences of the GBV survivors, male and female inhabitants of Barguna who were interviewed in the research.

> *GBV in coastal regions includes many things. Women face disasters, they have problems with their husbands and they have children to take care of.*
>
> *(Female, Barguna)*

Respondents were asked about their experiences and perceptions of the prevalence and nature of GBV in Barguna. Their responses suggest that GBV is a very common and chronic occurrence within their communities during non-disaster periods. Forms of GBV include child marriage, domestic violence, rape and sexual abuse, eve teasing (street harassment), dowry violence and psychological abuse. Of these, domestic violence against wives, child marriage and dowry violence are the most common.

Domestic violence against wives and its causes

Domestic violence against wives is the most common form of GBV, committed in Barguna in many households. Though it is a very sensitive and private issue to talk about, survivors were keen to describe their experiences and even showed their bruises during the interviews. Their experiences were difficult to hear, showing how vulnerable many girls and women in Barguna are in their own homes.'*I was beaten with a shirt, hands, several times for silly or simple reasons, one day he pushed me towards the wall which knocked me senseless*' (Female, Barguna). The research suggests that power relations, culture and tradition, and women's empowerment are the main influencing factors, creating an environment that enables this violence. However, the reasons given for wife abuse vary according to time, place and age of those targeted. Sometimes, there were no specific reasons mentioned, and women are simply beaten by their husbands. According to one victim-survivor, '*He beats me every day. Whenever he wants to beat, he starts. He uses bad words, he swears*' (Female, Barguna). Other interviews reveal simple reasons such as the wife not finishing cooking on time, or not completing the husband's orders or maybe he is just upset and becomes violent. '*He hit me with glass... I got a cut on my hand...I was pregnant that time. The reason was I spoke loudly*' (Female, Barguna).

As is clear from these accounts, the 'reasons' given by husbands for their abuse of wives are not real causes of the abuse, but rather excuses for abuse which will happen anyway. In Chapter 6, we discuss the causes of GBV which lie in patriarchal systems and cultures. However, there are many aggravating factors that make violence more frequent, more severe and which impede survivors' attempts to resist and overcome it. The most common of these factors mentioned by the respondents were dowries, poverty, extra-marital relationships and extra marriages, and traditions; these are described in more detail below.

Dowry

Dowry is the main reason mentioned by respondents for wife abuse. Nearly every marriage in Barguna involves some form of dowry violence. Dowry is the money the parents of the bride have to spend or the gifts they must give the groom and his family. The amount of money depends on many things. For example, the status of the groom – if he is educated, the dowry will be higher; or if the family of the groom is high status, the dowry will also be higher. On the other hand, if the bride has a dark complexion, has any disability or any social record that is perceived as negative, the dowry will also be higher. One respondent said,

> *I have dark complexion and I do a job outside my home, these are the main problems for my family to get a good marriage proposal for me, dowry money is higher...everyone (neighbours) talks about me negatively...I became a burden on the family.*
>
> *(Female, Barguna)*

A negative social record may involve many complex criteria. One of the respondents was a trafficking victim when she was only six or seven years old. Once she was older, her father had to find a larger amount of money to arrange her marriage. He is very poor and struggles to earn enough for his family's daily meal; however, he had to promise one lac (GBP 1,000) as dowry money. He could not find this amount in time, and as a result, the girl was beaten by her husband and his whole family:

> *They blamed me and say 'your father did not tell us that you were trafficked'. They also want full dowry money from my father...though my father informed them about trafficking before the marriage and spent a huge amount of money already...but could not fulfil their demands ...so they torture me. My father is very poor.*
>
> *(Female, Barguna)*

A negative social record may also be perceived when a girl's family live near a bazaar, or the father is a shop owner in the local bazaar and the girl works at her father's shop; or if the girl has been eve-teased or abused by local men; or the girl has been in a relationship with a boy or her neighbours simply suspect this. These criteria are very seriously considered during the marriage proposal and in deciding the amount of dowry money.

Besides these criteria, women are often beaten if the promised money is not paid on time, or in order to extract more dowry money. It seems a common phenomenon that the family who take dowry during marriage then become greedier, torturing women to get more money whenever there is a need for it:

> *He orders me to bring money from my parents but where will they get money? They are poor and do not have anything left to give him. He beats me for this.*
>
> (Female, Barguna)

> *The more educated the groom is the higher the dowry money is. She is tortured, beaten, forced to abort her child only for dowry money. She still does not want to leave her husband's house, to conserve the honour of her family.*
>
> (Female, Barguna – this respondent discussed the case of her daughter who experienced dowry violence)

Poverty

> *He beats me with whatever is near to his hand. Look at my bruises. These marks will never go away. He wants me to bring cigarettes for him, he wants me to bring money… where can I get money?*
>
> (Female, Barguna)

According to the Barguna residents, poverty is one of the main reasons for wife abuse. Whenever the family falls into poverty, violence against the wife increases. Loss of jobs or businesses, disasters, health problems and loan debts can create pressure leading to domestic violence. Loans exert a double-sided pressure on women. Sometimes, they are forced to take on loans, because small loans from private organisations are mostly issued in the wife's name, but many examples emerged in this research where husbands did not want to share the responsibility of paying instalments, stole the money from their wives or sold the item bought with the loan money and spent the proceeds. When these women try to claim the money back from their husbands, they are abused:

> *Wife beating and conflicts in families are mostly created because of loans taken from different organizations. When the instalments cannot be managed then both men and women in the family become tense and they start quarrelling. However, women are mostly beaten by their husbands.*
>
> (Female, Barguna)

> *He stole my rickshaw which I bought with the loan money and he sold it.*
>
> (Female, Barguna)

> *My husband cheated on me, he took loans from everyone he knew, used me to get rid of his creditors and humiliated me in crowds, accused me of negligence of my responsibilities in relation to loans.*
>
> (Female, Barguna)

Extra-marital relationships and extra marriages

Extra-marital relationships, adultery and extra marriages are also important reasons given by respondents for wife abuse:

> *If a man starts liking women who live outside home, would he be happy with the woman at home? He has beaten me from the very beginning and bruises are all over me. Those are very painful and I became sick.*
>
> *(Female, Barguna)*

The study reveals that this is a very common reason given for husband–wife conflicts in many families in Barguna. When men are in several relationships they want to avoid their wives, and/or they become violent. However, their behaviour is generally tolerated by society in the name of tradition.

In the name of tradition

Many men and some women in Barguna justify wife abuse in the name of tradition. They think that women need to be punished when they do not follow family norms or their husbands' orders. This belief that women should be controlled by men leads to social acceptance of spousal abuse, especially of women, which is also present in other cultures, as Linos et al. (2012) mention in their paper on Iraq. However, some women in Barguna are not only beaten by their husbands but also by their mothers-in-law and even other members of the husband's family. Many respondents mentioned this in their interviews:

> *All of them beat me, he (husband), his mother, brother, father all beat me.*
>
> *(Female, Barguna)*

> *He and his ex-wife have beaten me and evicted me from my own house.*
>
> *(Female, Barguna)*

Again, wife beatings are considered to be families' own matter that should be kept private. Most of the time, neighbours remain as passive bystanders, only helping to stop the violence when it is judged to become very severe, or if the woman seeks help from them.

Child marriage

Child marriage is also a form of GBV. Girls are married off at a very early age in Barguna, sometimes as young as ten years old – almost always to adult men. According to the respondents, poverty is considered to be the main reason for child marriage. However, there are also other reasons, as a respondent explains:

> *Parents feel the lack of security for their girls, and they do not believe in higher education for girls, so they arrange child marriages for their daughters.*
>
> *(Male, Barguna)*

Child marriage (below the age of 18 years) is illegal in Bangladesh. However, the parents are so determined in their plans that it becomes difficult for the government or NGO officials to prohibit them. Parents even make fake birth certificates to marry off their daughters at a very early age. The key informant, who worked on child marriage in a professional capacity, described the impacts of child marriage on victims and the difficulty they face in prohibiting it:

> *From my experiences I find that 80 to 90 % of child marriages are ending in divorce or abandoning the wife. These girls are also becoming the victims of dowry violence. We try our best to stop child marriages whenever we receive information that one is being planned. However, these parents are desperate, they proceed with their plan when they have made up their mind.*
>
> *(Female, key respondent, Barguna)*

> *Child marriage in Barguna is more common in fishermen's society. They move from one place to another in different seasons. They get married to young girls and when they leave the area, they leave the new wives too.*
>
> *(Male, Barguna)*

Child marriages are often followed by other forms of violence, such as wife beating and dowry violence:

> *Many times parents married off their daughters to save them from the local men who might abuse them, however these girls do not know what marriage is. They are often scolded, beaten by their husbands and in-laws.*
>
> *(Female, Barguna)*

Most of the time, these victims of child marriage cannot continue their education, which can be seen from the education statistics of Bangladesh. The number of girls attending primary school is satisfactory, but decreases dramatically after this level (BBS 2013b). In many cases, this is because girls are forcibly married. Often, during the marriage proposal the groom's family promise to allow the girl to continue her schooling, but like the other promises they make, they do not keep it: '*They did not let me continue my study though they promised*' (Female, Barguna). These young girls may also have their basic needs neglected by their husbands: '*My husband did not believe me, did not take care of me during my illness, I was not taken to the doctors*' (Female, Barguna).

Additional marriages and wife abandonment

Extra-marital relationships were named as a problem in many respondents' lives. It was reported that many women had been cheated by their husbands, who then abandoned their wives to remarry. This may also commonly involve violence. The behaviour could be noticed from the very beginning of the marriage:

> *He has extra-marital relationship. He liked his sister-in-law before my marriage. His family knew that but they forced him to marry me.*
>
> *(Female, Barguna)*

He eloped with a married woman when I was six months pregnant with my child. He remarried her without my consent.

(Female, Barguna)

We did not know that he was in a relationship with another woman before my marriage. Still he married me, and I then found that he was cheating. He started avoiding me after some time. He gave my phone number to some bad men and they now call me and disturb me.

(Female, Barguna)

My husband left me when I was at my father's house to give birth to our first child. He cheated and went to Dhaka and married another women.

(Female, Barguna)

When the husband marries again, he sometimes wants the first wife to stay in the same house as his new wife. This creates an embarrassing and sometimes vulnerable situation for both wives; often conflicts start, which may end with violence and crime. However, if the wife does not want to continue her marriage, she might be accused by her husband and by wider society too.

Rape, abuse and eve-teasing

Rape is mostly kept secret and hidden in Bangladeshi society. Most cases are not reported to any authority, whether the government or NGOs. However, the present research revealed several cases of rape and abuse in Barguna, including some where the victims were below 12 years old. Rape cases also occur in poor neighbourhoods, sometimes committed by richer members of society. In particular, in the fishermen's village, cases of rapes were higher. The stories are not detailed here as they are very sensitive and disturbing.

Rapes are sometimes conducted by the cousins or other members of the family.

(Male, Barguna)

Eve teasing (street harassment) is also a great concern for young girls and women, including married women, in Barguna. There are several cases of eve teasing reported; even one of the researchers of this study experienced this during fieldwork in Barguna. Respondents mentioned eve teasing and abuse several times in their interviews:

My husband left me… it is a violence to me. However, when people swear at me, say bad words to me, it also becomes violence to me. When I walk by the road they tease me, only because I am a woman and my husband left me. I have no faults except these.

(Female, Barguna)

Trafficking

During our fieldwork, we interviewed a trafficking victim-survivor. The interview reveals that she might have been saved from trafficking, but she has now become the victim of dowry violence, wife abuse and psychological abuse in her husband's house. She was pregnant at the age of 17 and did not receive any vaccinations during her pregnancy due to her husband's negligence. She was often beaten.

Other respondents talked about the commonness of trafficking, describing it as a hidden problem:

> *Trafficking is present in Barguna, however it is not easy to recognize. Trafficking is conducted with the permission of parents or victims themselves. These men or women leave Barguna for jobs offers by the traffickers and many of the victims do not return. We know presence of this violence but do not know more.*
>
> *(Male, Barguna)*

> *Even parents sell their daughters, or provoke them to get involved in prostitution. They agree to give their daughters to do jobs as housemaids. However, many of them know that their daughter might not be safe. Parents try to reduce their responsibilities for their daughters and the burden of poverty.*
>
> *(Female, Barguna)*

GBV and male survivors

There are some incidents of GBV mentioned in the interviews where men were victimised. According to some of the male respondents, the present laws against GBV are seen to work for women, who use these laws for their own purposes. For example, in one case we were told a woman filed a false case against her husband because she did not want to be with him. Besides, women may become oppressive when they bring money and household things from their father's house at the start of their marriage. A respondent explained, '*If a woman brings things (dowry or gifts) from her father's house, she might start teasing her husband and his family*' (Male, Barguna). The respondent called this mental torture, or psychological abuse, for the men.

Some women also become involved in extra-marital relationships and cheat on their husbands, especially during the husband's absence:

> *They have extra marital affairs and they cheat on husbands, especially married women have relationships with young men, and many of these women's husbands live abroad.*
>
> *(Male, Barguna)*

A few cases were reported where women start fighting with their husbands, throwing things and beating them. These incidents led to separation or divorce. However, our male respondent who was victimised by his wife said that he still wants to continue in his bad marriage, for his son and for the honour of his family.

A key informant who works in the Mahila Parishad mentioned that '*Male victims also come to us to get help and retrieve the marital relationships*' (Female, Barguna). In addition, we interviewed a male who was a victim-survivor of sexual abuse that involved unwanted touching when he was 10–11 years old. He reported that a middle-aged man who lived near their house often touched young boys like him, and he and his friends were the targets of his violence. He had not mentioned it to anyone except me in his entire life. This behaviour is very secret in this society, and very rarely talked about. However, the male respondent's courageous effort shows the presence of this violence and the vulnerability of young boys.

Nonetheless, compared with men, it is clear that women are the main targets of all types of GBV. The experience of GBV is very common and the severity of many cases is alarming.

2.5 GBV before, during and after disasters

It is already known that the number of incidents of GBV increases after tropical storms. Molin (2018) reported increased physical and sexual violence in the Philippines after typhoon Haiyan and typhoon Bohpa (GBV AoR 2013 in Molin 2018). As discussed so far, the present research has revealed that GBV occurs to an alarming extent in Barguna even during normal times of the year. However, GBV then worsens before cyclones hit (during the warning periods), during cyclone events and after cyclones (in the post-disaster period). Violence against women increases in incidence and also in its different forms. According to the respondents, common violence like domestic abuse, early and forced marriage, rape and other forms of abuse, cheating and wife abandonment also increase, thus further compromising the security of women after disasters.

Rape, sexual abuse and threats during disasters

The research reveals experiences of sexual abuse during disasters. It shows, firstly, that disasters such as cyclones create a context that enables more sexual harassment for Barguna women. From the previous section, we have come to know that verbal abuse is common in Barguna; however, sexual abuse and rape are also common. Based on respondent reports, its incidence appears to increase before, during and after cyclones, from the time of the weather warning to the post-disaster period. Many respondents were the victims of sexual harassment and described their experiences during the fieldwork in Barguna. However, no rape survivors agreed to be interviewed (or, no interviewees wished to divulge rape). The key informants informed the research team about rape incidents after disasters, but those affected did not want to talk about it. Therefore, the following sections reveal the prevalence of sexual harassment, which helps to build up a picture of GBV in disasters.

During the warning period before the cyclone hits

Weather warnings are disseminated by hand microphone, radio, TV and mobile phone networks before the cyclone hits the land. Many volunteers affiliated with government and local NGOs work to ensure the weather warning reaches every house. Following the warning advice, people who leave their houses mostly take refuge in cyclone shelters, a neighbour's stronger house or at nearby offices. Local people, mainly male, take part in the dissemination of warnings and advice before the cyclone, helping people to reach safe shelters. These activities, therefore, create the context for gatherings of unknown males in the neighbourhood. Some respondents reported that under the guise of volunteering, some men enter the locality and use the opportunity to harass women. Some of these women have been separated from their families or larger groups on the way to shelters. The men even check the empty villages to see if any women have been left behind, and attack them, including female-headed families. Instead of helping women seeking refuge from the cyclone, then, some men become perpetrators.

> *Sometimes people come to help the people to reach the safe shelters. However, some of them take the chance to rape, abuse the girls or women found alone at home.*
>
> (Male, Barguna)

During the disaster: GBV at cyclone shelters

Women's vulnerability to GBV continues when they reach cyclone shelters. It is important to say that many respondents reported no violence at the cyclone shelter, while some reported a lot of abuse. Differences in perceptions of GBV might have influenced these opinions. While unwanted touching and verbal abuses were common in crowded shelters, mentioned by many women and by male respondents who saw and protested the incidents, some female respondents did not say anything about it. Some see this male behaviour as normal; '*men are always like that, this is normal for men, they are not good*' (Female, Barguna). However, many women did express their fears and shared their experiences. Some of the memories from the recent cyclones were very fresh, and while they were being described, it was difficult to remain composed; rather, fear captured both of us – survivor and researcher.

Here, the respondent describes a case of harassment and threatening behaviour, which occurred during Cyclone Sidr:

> *There was a young girl who took shelter during Cyclone Sidr. Her mother was sick. Her mother left her at the cyclone shelter asking me to look after her. After some time, some men came and asked her to go with them to talk in private. The girl was afraid and did not want to go with them first. However, they forced her. I told them that if you want to talk with her let her mother come first and then you might talk with her. But they forced again and she had to go with them to talk. I was standing a little far from them. They proposed an illegal relationship and threatened her. She came back*

> *to me shivering. I also became scared. We had to leave the shelter when her mother returned to the shelter.*
>
> *(Female, Barguna)*

This incident shows how young women experience harassment while staying at the cyclone shelters. Mothers of girls mentioned several times that they were afraid to take refuge in shelters. Their experiences in previous cyclones include verbal abuse, unwanted touching, and men tearing their dresses in the crowds. While people push each other to enter into the cyclone shelters, some men use the opportunity to touch women intentionally.

> *The shelter is not safe for us. Young men come from 7 or 8 villages. They eve-tease [verbally harass] girls and young women. They try to touch or molest them. I feel frightened to stay in the shelters. I stay at my house rather than taking my teenage daughter to the shelters.*
>
> *(Female, Barguna)*

All such incidents work as a strong factor that keeps women and girls at their vulnerable houses, choosing to avoid cyclone shelters which then increases their risk during cyclones (Paul and Routray 2011).

Sexual harassment during the post-disaster period

After returning to their homes, survival in the post-disaster period becomes a great challenge for families. Most of the houses have been damaged, and food and water insecurity hit everyone in the affected area. According to a number of respondents, men from outside the village continue to roam around the damaged houses and disturb girls and women:

> *After the disaster, some young men roam around the area. They tease young girls, women. They sit here and there and pass time. This is a problem after disasters.*
>
> *(Female, Barguna)*

Women are also harassed outside their homes when they travel for healthcare or for collecting relief (Rezwana and Pain 2020). It can become very difficult for women to move around without chaperones, whereas men are busy with relief collections and some of them remain ignorant to what is going on for women.

Relief is very important for surviving during post-disaster periods, especially for the poorest inhabitants. However, there are many problems affecting relief distribution, such as mismanagement, insufficiency and corruption. The current research reveals that these problems are more intense for women, as they have to face the risk of GBV when they receive relief and recovery support.

One of the respondents, 'Muna', had a very bad experience. She is the main earning member of her family. Her husband left her and remarried, and he no longer takes any responsibility for her and his children. The reason for leaving her is, as she described it, '*I am old for him now at this age...*' Muna is 42 years old and her children are seven and four years old. Muna's age and marital status also became reasons for her to be targeted by a perpetrator. After a recent cyclone, Muna was trying every source to get relief for her family, asking everyone she knew to help her to survive the post-disaster period. She told us that she had managed to get some relief from different sources but it was not enough. One day, a man who was barely known to her told her to come to a place near the bazaar to get some relief. She was confused and did not fully believe him, but felt that she couldn't say no to him and had to take the chance. So, she took a chaperone, a known male, and went to the bazaar in the evening. She asked her chaperone to stay a little distance away from the building, not daring to let the man know that she didn't believe him and had brought someone else alone for safety. She entered the house alone, immediately felt that something was totally wrong and left immediately. There was no one in the house except the man, and there was no arrangement for relief distribution. She saved herself that day and returned home. However, she later got phone calls from the man, which she didn't answer. She said, '*I was too afraid to receive the calls and still I feel panicked about what could have happened to me*' (Female, Barguna).

According to the interviews, this is not a rare case of threatened GBV in relation to relief distribution; respondents mentioned several other cases too. One case just after the cyclone in 2016 helps further understand the context of GBV in Barguna. It took place in a crowded office where rice was being distributed among the poor. Not everyone was getting the rice, and so there was tension among the people waiting. 'Luna' was present in the office with her younger brother. While they were waiting for the service, they noticed that a woman had been sitting in the office from the early morning. No one was attending her and she was not in the queue. In the late afternoon, the woman was called by an official and he took her to an empty room upstairs. Everyone knew what was going to happen and some were laughing and mocking. This incident shows the vulnerability of this woman who needed the support very badly. She became a survivor of GBV in order to survive the poverty that followed the disaster.

Post-disaster poverty and domestic violence

Our informants reported that domestic violence against wives increases, in incidence and severity, after every cyclone in Barguna. Similar increases were noted in Tafe Province in Vanuatu in 2011 after two tropical cyclones (UNWOMEN 2014). One of the GBV victim-survivors in our study described how her husband became abusive after Cyclone Sidr and continued to this day:

> *My husband and my father are both very poor. Every cyclone intensifies poverty for them. My husband lost his fishing nets and other tools during Cyclone Sidr... he became violent. He started beating me. He beats me listening to his sisters. They complain about me and he beats me. One day I had to leave the house and took shelter to*

> *my aunt's house to save myself from his tortures. He brought me back. But, still he is abusive and beats me.*
>
> (*Female, Barguna*)

Intense poverty after disasters, the lack of jobs and money, and the loss of housing and possessions all create huge tension among family members. As one of the respondents described:

> *During disasters, the family might lose everything. Cattle, trees, crops, house... they become penniless. They might not be able to earn these resources again in their whole life. People become restless and feel lost. In these families, conflicts begin and wife beating increases. People even fought over receiving relief.*
>
> (*Male, Barguna*)

Poorer inhabitants do not have any savings for emergencies and need help after cyclones. However, there are huge issues of mismanagement reported in relief distribution. Not all of those affected receive help, and the amount of relief is often insufficient. These conditions are described by respondents as provoking some men to abuse their wives as a form of blackmail to extract money from their parents. However, as disasters strike everyone, often the wives' parents do not have any money to give. As a result, domestic abuse increases further. Respondents explained:

> *I could have brought a few things from my parents before the cyclone, but after the cyclone my parents also became helpless...They faced severe losses in the cyclone and still they haven't recovered from those losses.*
>
> (*Female, Barguna*)

> *Torture for dowry money increases after disasters. If I could bring some money then I would get relief from beating, and could stay in peace for some time.*
>
> (*Female, Barguna*)

When conflict and abuse between family members increase after disaster, both men and women may participate in these incidents. One respondent explains, '*People become restless and lose their patience in disasters, that's why more quarrels occur in families*' (Female, Barguna). Women may also get involved in family quarrels. In post-disaster periods, women need men's help to earn money, collect relief and repair houses, as it often takes time for women to regain their previous jobs due to fewer opportunities as well as their extra labour with cooking and taking care of family members in a damaged environment. However, men may lose their jobs and then intense poverty strikes, with some reported to the research as not making efforts to find new employment to cope with the emergency conditions, which increases family conflicts. This is another very common dimension of GBV reported in Barguna after disasters:

> *My observation is that whenever men face a lack of money, conflicts between men and women increase...especially after disasters. Fishermen in this area become jobless after cyclones. They do not know or are interested in any other work. All day they roam around, play cards or gossip and spend leisure time. On these days whenever they return home women start questioning them. Women are helpless at home, waiting for men to bring food or other necessary groceries...so they start accusing. Then conflicts begin and wife beating increases.*
>
> *(Male, Barguna)*

Child and forced marriages after disasters

Intense poverty and the lack of security after disasters increase the number of child- and forced marriages in Barguna. Goulds (2013) reports similar patterns affecting young girls in nine countries. Girls may be forced into marriage at a very young age; one reported to our research was only 13 years old. She was studying in class 5 at school and forced into marriage just after Cyclone Sidr. She talked about her experience during the interview, highlighting the reasons for and consequences of this form of abuse:

> *I was married just within two months after Cyclone Sidr. My father was very poor and the cyclone intensified his poverty. The cyclone took everything we had. Again, my father became afraid of insecurity and being humiliated by the society. He thought people would say bad things about me. So, he arranged a marriage for me as early as possible. However, coming to my husband's house I found that poverty and other problems are more intense here. At my father's house whatever he earned I could have eaten, but now I get my meal after everyone else. Besides, my husband beats me every day. I tolerate all this for my honour and my children, and I have no place to go.*
>
> *(Female, Barguna)*

According to the victims of child marriage, their parents would have said no to a proposal for a young girl before the cyclone, but after the disaster, they agreed, seemingly without thinking about the consequences, and their daughters were forcibly married. An opportunity arose to talk with a parent who forced her daughter's marriage at a very early age. She explained the reasons behind her decision:

> *I had to say yes to the marriage proposal for my daughter after Cyclone Sidr because we were in intense poverty, no food, no proper shelter. They (groom's family) said they would provide all the marriage expenditure. I had to say yes.*
>
> *(Female, Barguna)*

However, partly as these marriages are arranged quickly, in many cases, fraud takes place. Grooms hide their age, their economic circumstances and even previous or existing marriages. Parents may not have the time or interest to check the information given to them about grooms. As a result, girls become subject to even more abuse. As one respondent explains:

> *I was married just after the cyclone Sidr. Poverty increased after the cyclone and my parents did not make checks on the man. The man was old, married and had a wife and two children already. He lied and cheated me.*
>
> *(Female, Barguna)*

This young woman had to face her husband's first wife and was physically abused by her. Another important reason for forced child marriage after disasters is that rich men exploit disasters as an opportunity to prey on girls from poor families. For the reasons explained above, and because the families do not want to take a risk keeping their young girl at their damaged, unsafe house, and prefer to arrange a marriage for her. However, the research reveals that the most common underlying reason for child marriage is that whenever a family falls into crisis, girls are thought of as a burden to the family. The costs of keeping the girl in the household are offloaded by the family through forced marriage.

In a related case revealed in the present study, a forced marriage had taken place in the study area after paying a 'ransom' after Cyclone Sidr. The respondent described:

> *In our village, a woman took a loan from a man after Cyclone Sidr and could not pay it back within the time stated. So, he forcefully married her. This man has another wife and two daughters already.*
>
> *(Female, Barguna)*

Again, this illustrates the very disempowered condition of women after disasters, and highlights prevailing social attitudes that are accepting of early and forced marriage in Barguna. Although this is a common form of violence, the survivors' descriptions reveal that members of society do not protest it.

Social crisis and abuse in the post-disaster period

In post-disaster periods, the severity of crimes and violence increases generally, influenced by existing patterns of GBV in the locality. One alarming finding from this research is that other family members (such as first/other wives, mothers-in-law or sisters-in-law) sometimes also participate in abuse besides their husbands, sons and brothers, making the situation even worse for women. Sometimes the husband and his family force women to collect relief. As relief distribution policies often emphasise female disaster survivors, women get priority. However, the rural culture of Bangladesh does not support women from different socio-economic groups joining public relief queues. Additionally, we have seen in the previous sections the added dangers for women in relation to relief collections. However, families sometimes become forceful and women may even be beaten if they do not go out to collect relief.

The present study also reveals that their family members sometimes deprive women of their rights. For example, after women collect money as relief, their husbands may take it. One of the respondents was thrown out of her own house by her husband and another wife, after she collected relief. The case is described as follows.

Aklima's father forced her marriage to an older divorced man. Like other young girls, Aklima could not say no. However, after a few years of marriage, her husband left her and remarried. Aklima was trying to survive alone by doing different informal jobs. However, when Cyclone Sidr hit Barguna, like other poor women she lost everything. Fortunately, she received a small house as relief. She made a pond, and made the house liveable with other people's help. After some days, her husband returned to her. He brought his other wife and both of them started abusing her. One day they beat her severely and evicted her from the house. Aklima was crying during her interview and saying that if she knew this would happen to her she would have not spent every penny and effort into that house. She is now living with her poor parents. She is in a very weak position, physically, socially and economically, and cannot take any action to get back what is rightfully hers.

Other cases are reported of looting and taking relief from poor women, violence that is mostly committed by husbands and their families, and sometimes neighbours. On some occasions, husbands may use their wives' names to get loans more easily.

Another type of abuse is found in the study area after cyclones. During the post-disaster period, a scarcity of informal jobs becomes a great problem for poor families. In this situation, male family members often leave the area to get a job in the big cities. In many cases, these men send money for the first few months, and then stop contacting their families. Women are abandoned with their children in the disaster-affected area, amid immense poverty and insecurity.

2.6 The impacts of GBV: health, socio-economic position and vulnerability to future disasters

The research reveals various impacts of GBV on survivors, which are often both multiple and severe and endure well past the disaster and post-disaster periods. This is the case both for direct impacts (physical injuries and psychological impacts, discussed in more detail in Chapter 6), and indirect impacts such as changes in economic conditions and social security.

Physical and mental health impacts of physical violence

Physical injuries mostly occurred due to domestic abuse, particularly against wives. The survivors face a range of injuries, from minor to severe, due to beating by their husbands and/or other family members. This type of abuse, as reported in this chapter, increases in disasters, and it often has severe ongoing impacts on survivors. According to the respondents, they were mostly beaten with hands or items to hand such as wooden or bamboo sticks, brooms, mugs, shirts, etc.

> *He threw a glass towards me...it hit. I was pregnant at that time.*
>
> (Female, Barguna)

Respondents reported sometimes being hit on the face, which can cause more pain and sometimes lifelong disabilities. Some were beaten until they became unconscious. Survivors of this kind of violence had broken limbs, lost eyes or developed hearing problems:

> *He hit me on my eyes three times. Blood got clotted in my eyes. I cannot see well with this eye. I cannot even thread a needle.*
>
> (Female, Barguna)

> *One of my fingers was broken as he hit me so hard one day. He always hit me.*
>
> (Female, Barguna)

> *He hit me on the face so hard that I cannot hear properly now. He started punching my face ...again and again...I became senseless. I have got a bone injury. He did not take me to the doctors, my father did.*
>
> (Female, Barguna)

From some women's descriptions, it also appeared that there may be evidence of chronic traumatic encephalopathy or 'punch-drunk syndrome' which is common among ex-boxers as well as survivors of domestic violence (Casper and O'Donnell 2020), but in Barguna women, it had not been diagnosed. Survivors mostly have to rely on their husbands to visit the doctors, as female patients need a chaperone, permission from their husbands and above all, money for the treatments. Husbands who have been violent rarely take their wives to the doctors, so their injuries remain untreated. In very few cases, a parent takes their daughter to the healthcare centre. As a result, they experience longer pain and suffering, and many internal injuries remain undiagnosed.

> *He beats me whenever he wants. I became sick due to these tortures. One day he hit me on my ear, it started bleeding...I cannot hear well now.*
>
> (Female, Barguna)

Researcher: 'Did you get any treatment?'

> Respondent: *How can I go to the doctor? Who abuses you, is he going to pay for it? He says "die...if you die I would be happy". How can I die? I have two little daughters.*
>
> (Female, Barguna)

Especially during disasters such as cyclones, there is no opportunity to visit the doctor. When injuries caused directly by the disaster do not get treated, due to damaged transport systems and healthcare centres, and lack of money, it is clear that injuries from GBV are even less likely to be treated. However, it is alarming that domestic violence also increases during the post-disaster period. The resulting physical injuries, along with increased household work in damaged environments and disruption and trauma from the disaster, create a very difficult period of survival for the women affected. The health impacts of rape could not be discussed in this research, as rape survivors did not want to talk about it during the interviews. One trafficking victim described the physical health problems she still faced from injections by the traffickers during the periods of captivity. The substances she was injected with damaged her ability to remember events, and she was unable to complete her primary education afterwards. Trafficking also had a severe psychological impact on her. She said she could never feel like a normal girl and had lost her childhood. Being the daughter of a poor family, and living in a conservative society, she never received proper treatment for these physical and psychological problems. The significant issue of psychological trauma after GBV during disasters is discussed in more detail in this research in Bangladesh in Chapter 6 and, in a UK context, in Chapter 4.

GBV and changes in socio-economic status

Another significant impact of GBV on survivors' lives, especially during disasters, are changes in economic and social status. These impacts are also life-changing, as well as being closely connected to mental health. Most women in Barguna do not work outside the home to earn money, but are fully engaged in household work. However, extreme poverty may lead them to earn from the informal sector, except for the few educated women who choose to take formal jobs. The respondents in the present research mostly belong to either lower middle class or extremely poor economic classes, and mostly do not earn money except those who are divorced, separated or abandoned by their husbands. These women frequently become the poorest of the poor after losing economic support from their husbands, as we have described earlier in this chapter. Before the violence, their world was within their home, but afterwards, they often become day labourers or maids in other houses. Even then, many of them cannot earn enough for every meal and may starve for days.

Divorce, abandonment and separation also remove shelter, education, healthcare and food security from women and their children. Almost all survivors in the research mentioned these socio-economic changes. Rahima was one of them. Explaining how she became vulnerable after being evicted from her husband's house, she expressed her regrets and sorrow:

> *He started torturing me when he was planning to bring his ex-wife back. If I knew that I would not have collected and stored all those food grains. They (husband and wife) beat me and evicted me from the house... They did not give even one kilo of rice to my children, whereas I have left one big drum of rice in that house. I am struggling now*

> *with my poor parents. I have to work at another's house. It is so tough to earn a meal for my children. He does not even take responsibility for the children.*
>
> (*Female, Barguna*)

Like Rahima, many survivors had ceased their education due to child marriage, and could not access proper healthcare or complete treatment to cope with the economic changes. According to Masson et al. (2016), child marriage during post-disaster periods might initially enhance the coping capacities of the family, but girls who are forcibly married often have to cease their education and generally suffer from lower status within their families, which has psychological impacts on them.

Survivors of GBV become more vulnerable still if they have taken out a loan. One man was reported to have filed a false case against his wife and left the area:

> *They (husband and his family) filed a false case against us (me, my father and mother). He left me in debt and with this case hanging over us. My father became penniless and I am also helpless. I have to be dependent on donations.*
>
> (*Female, Barguna*)

Such women face severe problems during disasters. They do not get enough relief due to the absence of their husbands. Halima's case illustrates this point. She was left by her husband just before Cyclone Sidr and left to suffer the disaster with her one-month-old son. She lost everything in the cyclone. Halima explained, *'I stayed at someone's house without a pillow, bed or mosquito net with my one month old son. It was a horrible time. I suffered a lot'* (Female, Barguna). We heard also about the insecurity a single woman faces with regard to relief collection (discussed earlier).

This spiral into poverty for women and girls, compounded by GBV and made worse still in times of disaster, also has the effect of heightening their vulnerability to future disasters. There is thus a cyclical relationship between poverty, women's status in society, GBV and disasters such as cyclones.

2.7 Conclusion: women, the ultimate victim-survivors of GBV and disasters

This chapter has shown that cyclones intensify GBV in Barguna, Bangladesh. Although GBV is a severe and endemic problem at any time, it heightens before cyclones hit (during the warning period), during cyclone events, and after cyclones (in the post-disaster period). Given that this region is hit by cyclones quite frequently, and the length of time in which communities are still recovering from disasters, especially the poorest inhabitants, this framework of 'before, during and after' sometimes makes little sense.

Because of their gender identity and roles, women therefore face double impacts from any disaster. Along with disaster-induced injuries, the physical and psychological impacts of GBV do not always receive proper treatment, leaving survivors with long-term pain and health problems. Their economic circumstances often worsen after disasters as well: living standards drop, and food, health and education insecurities increase due to forced marriage, wife abandonment and other forms of GBV. While abused women are extremely vulnerable within marriage, separation and divorce may bring new forms of abuse and severe challenges to living conditions. Unprotected and unsheltered in a conservative patriarchal society, women sometimes experience several types of violence. In disasters, these conditions become severe. These findings show with stark clarity the interrelationships of geographical location, culture and gender with GBV and disasters, and their complex and detrimental impacts on women's lives.

References

Ahmed, S. M., Hossain, M. A., Chowdhury, A. M. R., & Bhuiya, A. U. (2011). "The health workforce crisis in Bangladesh: shortage, inappropriate skill-mix and inequitable distribution." *Human Resource for Health* 9(3): 1–7. www.human-resources-health.com/content/9/1/3

ASK (2021). "Violence Against Women-Rape: January-September 2020." www.askbd.org, accessed on 05/05/2021.

Banglapedia (2015). www.en.banglapedia.org, accessed on 22/10/2016.

BBS (2013a). Gender Statistics of Bangladesh 2012. Bangladesh Bureau of Statistics, Ministry of Planning, Government of the People's Republic of Bangladesh.

BBS (2013b). Bangladesh Population and Housing Census 2022, Community Report: Barguna, Bangladesh Bureau of Statistics, Ministry of Planning, Government of the People's Republic of Bangladesh.

BBS (2016). Report on Violence Against Women (VAW) Survey 2015. Bangladesh Bureau of Statistics, Ministry of Planning, Government of the People's Republic of Bangladesh.

Bradshaw, S. & Fordham, M. (2013). *Women, Girls and Disasters: A Review for DFID*, August. https://assets.publishing.service.gov.uk/government/uploads/system/uploads/attachment_data/file/844489/withdrawn-women-girls-disasters.pdf, accessed on 18/05/2021.

Casper, S. T. & O'Donnell, K. (2020). "The punch-drunk boxer and the battered wife: gender and brain injury research." *Social Science & Medicine* 245: 112688.

GBV AoR (Gender-based Violence Area of Responsibility) (2013). *Preventing Gender-Based Violence after Typhoon Yolanda; Responding to Survivors' Needs*. http://gbvaor.net/wp-content/uploads/2015/02/Preventing-Genderbased-Violence-After-Typhoon-Yolanda-Responding-toSurvivors-needs.pdf, accessed on 2/03/2017.

Goulds, S. (2013). In double jeopardy: adolescent girls and disasters, Commonwealth Education Partnerships. http://plan-international.org/girls/reports-and-publications, accessed on 11/06/2020.

Health Bulletin (2013). *Health Bulletin 2013, District: Barguna*. Ministry of Health and Family Welfare, Barguna Civil Surgeon Office, Barguna.

Hossain, T. K. & Suman, R. S. (2013). "Violence against women: nature, causes and dimensions in contemporary Bangladesh". *Bangladesh e-Journal of Sociology* 10(1): 79–91.

Linos, N., Khawaja, M., & Kaplan, R. L. (2012). "Women's acceptance of spousal abuse in Iraq: prevalence rates and the role of female empowerment characteristics." *Journal of Family Violence* 27: 625–633.

Loayza, N., Olaberria, E., Rigolini, J., & Christiaensen, L. (2009). Natural disasters and growth: Going beyond the averages. World Bank Policy Research Working Paper (4980).

Manusher Jonno Foundation (2021). Violence against Women and Children: COVID 19, a telephone survey, June 2020. http://www.manusherjonno.org, accessed on 06/05/2021.

Masson, V. L., Lim, S., Budimir, M., & Podboj, J. S. (2016). *Disasters and Violence against Women and Girls. Can Disasters Shake Social Norms and Power Relations?* London, Overseas Development Institute.

Matin, N. & Taher, M. (2001). "The changing emphasis of disasters in Bangladesh NGOs." *Disasters* 25(3): 227–239.

Molin, J. (2018). Preventing gender-based violence post disasters: building the capacity of humanitarian actors in the Philippines to engage with men and boys to reduce the risks of perpetration of violence. Unpublished thesis, Uppsala University, Sweden.

National Plan for Disaster Management (2010). *Government of People's Republic Bangladesh 2010–2015.* Dhaka, Disaster Management Bureau.

Paul, S. K. & Routray, J. K. (2011). "Household response to cyclone and induced surge in coastal Bangladesh: coping strategies and explanatory variables." *Natural Hazards* 57(2): 477–499.

Rahman, M. (2013). "Climate change, disaster and gender vulnerability: a study on two divisions of Bangladesh." *American Journal of Human Ecology* 2(2): 72–82.

Rashid, S. F. & Michaud, S. (2000). "Female adolescents and their sexuality: notions of honour, shame, purity and pollution during the floods." *Disasters* 24(1): 54–70.

Rezwana, N. (2018). *Disasters, Gender and Healthcare Access: Women in Coastal Bangladesh.* Oxon, UK, Routledge, Taylor& Francis.

Rezwana, N. & Pain, R. (2020). "Gender-based violence before, during and after cyclones: slow violence and layered disasters." *Disasters* 45(4): 741–761.

Sultana, F. (2009). "Fluid lives: subjectivities, gender and water in rural Bangladesh." *Gender, Place and Culture* 16(4): 427–444.

Sultana, F. (2010). "Living in hazardous waterscapes: gendered vulnerabilities and experiences of floods and disasters." *Environmental Hazards* 9(1): 43–53.

UNICEF (2021). Ending child marriage: towards evolution of social behaviour. https://www.unicef.org/bangladesh/en/ending-child-marriage, accessed on 10/11/2021.

UNWOMEN (2014). Climate change, disasters and gender-based violence in the pacific. www.asiapacific.unwomen.org, accessed on 20/03/2021.

World Bank (2016). Bangladesh: building resilience to climate change. The World Bank website. What we do: projects and operations. https://www.worldbank.org/en/results/2016/10/07/bangladesh-building-resilience-to-climate-change, accessed on 18/05/2021.

World Bank (2020). COVID-19 to add as many as 150 million extreme poor by 2021. https://www.worldbank.org/en/news/press-release/2020/10/07/covid-19-to-add-as-many-as-150-million-extreme-poor-by-2021, accessed on 28/05/2021.

3

THE IMPACTS OF DISASTERS AND GENDER-BASED VIOLENCE

Flooding in Rangpur, Bangladesh

Nahid Rezwana

3.1 Introduction: studying GBV on the banks of River Jamunashawari, Rangpur, Bangladesh

The previous chapter illustrated the particular vulnerability of women when it comes to cyclones and the ways in which gender-based violence (GBV) increases in its forms, frequency and severity when disasters strike. This chapter presents the second case study in this book through examining disasters caused by river flooding in Rangpur, another region of Bangladesh. Drawing on quantitative and qualitative research in a village that has been badly affected by flooding, it examines the relationships between this form of disaster and GBV. Again, the chapter's analysis comes to focus on the interrelationships between gender, culture and this form of disaster in this location. It also explores women's empowerment and how far this protects women from GBV, seeking to explore the cultural, social and economic causes underlying patterns of GBV.

Bangladesh is a riverine country where flooding is a natural phenomenon and one that is profoundly affected by human activity. Every year, the country experiences floods during the rainy season; however, the severity of flooding differs depending on the duration and river water level. Amongst its long history of floods, Bangladesh experienced severe flooding in the years 1987, 1988, 1993, 1998, 2000 (Banglapedia 2015) and recently in 2017. During the flood in 2017, 31 out of 64 districts of the country were affected, 121 people died, 238,843 people were seriously affected and 6,536,509 people were partially affected. About 15,529 hectares of agricultural land were fully inundated and 562,594 hectares were partially inundated during this flood (Philip et al. 2019).

Rangpur, an inland district in the far north of Bangladesh, was one of the most affected districts in the country; 8 upazilas, 61 unions and 487 villages within the district were flooded in 2017. About 11,714 people were displaced in the disaster,

DOI: 10.4324/9781003089780-3

and 165 km of roads and 38,815 hectares of croplands were damaged (NIRAPAD 2017). Kashiganj, in the Madhupur union of Badarganj upazila, is among the most severely affected villages in Rangpur. It is located on the bank of river Jamunashawari, which usually floods in rainy seasons. However, in 2017, the floodwater stayed longer than usual, causing a lot of damage to crops, cattle and houses. The whole village was underwater for more than a month. Maintaining daily routines became a struggle for residents. Most kitchens, sources of drinking water and sanitation systems were damaged, and movement became impossible as the roads of the village were underwater. GBVs like domestic violence, child marriage and rape were already high in the Rangpur division; it ranks second highest for physical violence against women (55.3% of women) among the divisions of Bangladesh (BBS 2016). The research reported here was conducted in 2019 in Kashiganj village (Map 1), aiming to reveal the impacts of flooding on GBV. This chapter will discuss the impacts of these disasters, the reasons behind the increasing incidence of GBV during flooding, and the interrelationships of GBV with gender, culture and disasters.

3.2 Background: the concurrence of flooding and GBV

Previous research on flooding in a number of countries shows that an increase in GBV during and after the event is a common crisis in flood-hit regions. GBV becomes a second disaster for the survivors.

For example, flooding is common in Pakistan, a disaster-prone country that ranks eighth on the Long-Term Climate Risk Index (Eckstein et al. 2018). A higher baseline rate of GBV is also a great social problem for this nation. Women experience different types of GBV in non-disaster times, according to a report 'Tracking numbers: State of violence against women and children in Pakistan' (DAWN 2021). This reveals that Pakistan scores 164th among 167 countries in the Women, Peace and Security Index 2019/20. The province of Sindh ranked second in child abuse in 2020 compared to the country's other regions. According to Memon (2020), Pakistani women are generally disadvantaged due to social norms, roles and responsibilities. She conducted a qualitative study in the flood-prone areas of Sindh to examine the association between 'natural' disasters and violence against women. Interviews conducted among women living in the flood settlement camps revealed that most had experienced different physical and emotional violence, mostly committed by partners but including incidents perpetrated by strangers. The rate of violence increases when women are displaced from their homes and living in temporary shelters during the post-flood periods (Chapter 2 reported a similar pattern in Barguna, Bangladesh, for women fleeing cyclones). GBV becomes both a threat to women survivors and to the effective implementation of disaster plans (Memon 2020).

The risk of GBV against women and girls also increased among IDPs (Internally Displaced People) during the Afar flood in Ethiopia in 2020. During this severe flood, 240,000 people were affected and 72,282 people were displaced. The loss of livelihoods led to migration and placed adolescent girls and women at greater risk of child labour, domestic violence and other forms of GBV. Emotional stress, trauma, and loss of socially enforced discipline worsened family conflicts. Women and girls living in the IDP centres also faced a lack of separate latrines and washing facilities, which put them at greater risk of GBV and health problems (CARE 2020).

Meanwhile, in September 2020, Sudan experienced heavy rainfall which caused the worst flooding in Darfur in over a century. Nearly 121,935 people were affected. The Darfur Women Action Group (DWAG) expressed their concern about the increased risk of GBV against women who left their community due to flooding. Women who were living in temporary flood shelters became vulnerable to sexual and physical violence due to the lack of safety and protective measures. Besides, Darfur has a history of using rape and GBV against women and girls as a strategy during the war in 2003, and unequal gendered power relations that intensify discrimination and violence against them during non-disaster periods. Within this context, natural events like floods aggravate these incidents, including domestic abuse, rape, sexual harassment and sex trafficking. However, the exact number of incidents remains unknown as various factors such as social stigma, cultural norms and a lack of accountability for perpetrators prevent women from reporting violence against them (DWAG 2020).

A rise in domestic violence and child marriage was also observed among disaster-affected people in Indonesia, Laos and the Philippines. A study conducted among 1,800 people impacted by storms and flooding revealed that many were worried about overcrowding in humanitarian shelters, a lack of support, and trafficking during disasters (Schlein 2018). Another study, identified increases in GBV including domestic violence, economic and psychological abuse, forced and child marriages, and the presence of other types of GBV such as rape, trafficking of girls, kidnapping and forced prostitution among the flood-affected Benin population after a severe flood in 2010 (CARE 2011).

During the monsoon flooding in 2020 in Bangladesh, GiHA (2020) found significantly higher rates of around 52% of psychological problems among women related to flooding. Rising rates of child marriage were also found due to the prolonged flood along with the Covid-19 pandemic. According to the Bangladesh Peace Observatory trend analysis, domestic and dowry-related violence increased 6% during the first six months of 2020, largely due to Covid-19, and which potentially led to further violence during the flooding crisis. Besides, although flood shelters accommodated around 25,377 women, the safety provisions in shelters were inadequate – such as separate toilets for women, adequate lighting, etc. – which put women and girls at further risk (GiHA 2020). Azad and his team also found GBV against women in the flood-prone region of Northern Bangladesh. Based in flood-prone Sirajganj District, the study revealed that 35% of women experienced harassment during and after flooding. About 90% of women were harassed (psychologically, physically or sexually) by their husbands, while 39% were beaten by them. Besides domestic abuse, 83% of women faced other types of GBV (physical violence 23% and sexual harassment 25%) while working outside home during the post-flood periods (Azad et al. 2013). Nasreen, in her research, also reported an increased number of GBV against women during and post-flood periods in Bangladesh (Nasreen 2008). Though flooding is a regular natural event in Bangladesh, however, the literature emphasises that detailed statistics on GBV during floods are rarely available in this country. In particular, data on different types of GBV during flooding, and accompanying factors such as geographical location, demographic and socio-economic conditions and gender relations are unavailable. There are few research studies overall, especially qualitative research. However, and as we asserted in Chapter 1, understanding the impacts of flooding on GBV requires in-depth contextual analysis of local culture and norms, and social attitudes to gender, gender identity and gender roles.

3.3 Aims and methodology

Given the importance of this understanding, the research reported in this chapter applied a multi-strategy design to collect both quantitative and qualitative data, in order to investigate the relationships between flooding and GBV in Kashiganj village. The research questions were as follows:

1. What are GBV conditions in the study area during non-disaster periods?
2. What are the forms and nature of GBV, during and after flood events?
3. What are the relationships between GBV, socio-economic status, education and employment?

A multi-strategy research design uses fixed and flexible design strategies and both quantitative and qualitative methods (Robson and McCartan 2016), which allows triangulation and increases the strength and richness of the data. The study was conducted for sequential explanatory design, first applying quantitative methods to collect data from the field, followed by the design of qualitative methods. The total number of households in Kashiganj village is 430, with a total population of 1,774 (BBS 2013). The sample size of 204 households was calculated using a statistical equation. These households were surveyed with a structured questionnaire in 2019, selecting every second house in the village to assure a random sample. The ratio of males to females was even among the respondents. The aim of this survey was to collect data on socio-economic conditions, education and healthcare facilities, and GBV conditions during both normal times and disaster periods in Kashiganj village.

Findings from the quantitative data analysis led to the second stage of the field study. Qualitative methods – in-depth interviews, focus group discussions, key informant interviews and observations – were used to explain and interpret the findings of the quantitative study, and to explore women's experiences of GBV and disasters in detail. Qualitative methods have 'an unrivalled capacity to constitute compelling arguments about how things work in particular contexts' and are 'capable of producing very well-founded cross-contextual generalities' (Mason 2002: 1), which was very helpful in this research in understanding the contexts of GBV and its consequences for survivors. Eighteen in-depth interviews were conducted among the inhabitants (male and female) applying a theoretical sampling method. Additionally, five key informant interviews were conducted, with a local school teacher, a housewife, a former local government committee member and two businessmen. Finally, observation was used during several long periods of fieldwork, which provided good opportunities to observe people's lives close at hand. The fieldwork was conducted between 2019 to 2020. This multi-strategy design helped to understand the contextual nature of GBV, and its complex relations to socio-cultural attitudes, norms and traditions.

3.4 Study area: demographic and socio-economic characteristics of Kashiganj village

Like many typical villages of Bangladesh, Kashiganj is located on the bank of a river (Figure 3.1); it has crop fields, muddy village roads, clustered houses, cattle shades and barns. However, a brickfield at the entrance to the village, consisting of newly

The Kashiganj village and Jamunashawari River

FIGURE 3.1 Kashiganj village

made large holes in the paddy fields (as soils are sold for making bricks) and tree leaves dusted with materials from the brickfield are special characteristics of this village. There is a growing number of half-*pucca/pucca* houses and the use of modern technology (mobile phones, internet connections and televisions), along with the signs of poverty in other sectors (education, health and other infrastructures such as roads and bridges); all this is juxtaposed, raising questions about the changing nature of the culture and socio-economic conditions in the village. The survey and interviews helped to gain insight into residents' lives and their perceptions of gender, GBV, disasters and socio-economic conditions.

Demographic characteristics (age, sex, education and occupation)

The inhabitants who participated in the survey were from different age groups: 26% are 16–25 years, 27% are 25–36 years, 25% are 36–45 years, 11.3% are 46–55 years and 10.7% were over 55 years. An equal number of male and female respondents participated in the survey. However, there are gender differences in education level and occupational status. Education level is generally very low among the inhabitants overall, with 54.5% of women not completing primary school education compared to 45.7% of men. Only 22.3% of men have completed SSC (Secondary School Examination) compared to 19.8% of women, and very few of them did a higher degree. Men are mostly involved in agricultural sectors (38%), business (36%), working as a day labourer (15%), in the service sector (5.26%) or unemployed (5.64%). There are a growing number of small businessmen who run shops in the bazaar, supply agricultural products, etc. However, 89% of women of this village are homemakers, and the others work as day labourers or housemaids. Very few (two or three) work as school/madrasha teachers, and one has a small street-food shop in the village. However, it should be mentioned that many women work as seasonal day labourers, sometimes help in each other's houses or have small businesses, but they did not mention this during the survey. The in-depth interviews revealed their participation in family income generation. The reasons for not mentioning these jobs in the surveys include thinking that a part-time job is not worth mentioning as a source of family income; being afraid of humiliation by saying it out loud, as lower-income jobs for women create negative impressions of them in the eyes of society; and because they took these jobs as one of their household responsibilities.

Household income and poverty

Occupational status along with cropland ownership has a direct impact on income and poverty levels among the residents. Kashiganj village has a large disparity in land ownership, with most of the croplands of the village belonging to four or five families while the other families share the rest. Most of the families have a homestead and small pieces of cropland which cannot provide sufficient crops for them. Table 3.1 shows that 84% of the families in the village are living below the poverty level.

TABLE 3.1 Distribution of income in Kashiganj village, Bangladesh

Income (Per person per day BDT)	*Number of families*	*Percentage (%)*	
0–160	172	84	Below poverty level: $ 1.90 PPP (Purchasing Power Parity a day: The World Bank 2020)
161–260	27	13	
261+	5	3	
Total	204	100	

Source: Author's field survey 2019

Repeated flooding is also a major cause of poverty in this village. The 89% of families who depend on the agriculture sector, small businesses or daily income as labourers remain jobless for one or two months during the rainy season and floods. In addition, every year there is damage from disasters. During the floods in 2017, 72% of families lost crops, 32.8% had their houses damaged, and 18% lost cattle, all of which create economic pressure. Affected families have to take out loans with interest from NGOs, in order to reconstruct their damaged houses and buy seeds for crop production. They remain in the poverty cycle long term due to the regularity of the floods.

Poor families rear cattle and plant trees to earn extra money. 75.5% of families have cattle or poultry, but on a very small scale – each family may have one or two cows or goats and a few chickens or ducks. However, flood water lying for a month or more always creates a major challenge for families to save their animals during these times of crisis. Many families have to sell their animals at a very low price during floods to get money for post-disaster economic recovery, or due to a lack of dry places to keep the animals safe. This is a very common phenomenon during flooding in other affected villages in Bangladesh (Rahman 2019).

The housing type of a family in the Kashiganj village has a strong relationship with family income. 39% of the families live in *kutcha* houses, which are made of mud, straw, bamboo, wood and tin. 41% have a *half-pucca* house. These types of houses have cemented floors, brick walls and tin roofs. Often only one or two main rooms in each house are *half-pucca*. These families take bricks from the nearby brickfields as loans and work as day labourers in the factory to repay the loan, and build these rooms over two or three years. Making *pucca* rooms is a flooding coping strategy. People prioritise this strategy to save their family members and cattle from the yearly floods, as they stay in the *half-pucca* rooms during the floods. Only 20% of families have full-*pucca* houses in this village.

3.5 Socio-cultural and gendered aspects of GBV in village life

As well as the risks posed by poverty and disasters, the presence of GBV is also a great concern in Kashiganj village. According to 69% of the inhabitants, various types of GBV (child marriage, dowry violence, domestic violence, sexual harassment and abuse) are major social problems in this village.

Child marriage, dowry and culture

Child marriage, which is very common in Kashiganj village, is considered a form of GBV as children cannot consent to marriage or sex. The study reveals that 81% of families in this village have cases of child marriage. Both girls and boys may be victims of child marriage; however, it is much more common for girls, with 78.6% of the women in Kashiganj saying they were married before the age of 18 years, while 40.1% of them were married before 15 years. This is much higher than the national average of 59% (UNICEF 2018 in MacQuarrie and Juan 2019). Among men, 35.7% had been married early, before they were 21 (Table 3.2).

The reasons behind child marriage are numerous and complex; however, it is commonly influenced by poverty, lack of awareness of the harm it causes and cultural customs. The amount of dowry money that girls' parents have to pay to the groom is one of the main reasons for child marriage, as dowry increases with the age of the bride. Here it should be mentioned that all of the families in the village have to pay dowry money irrespective of their daughter's higher education or income status. Dowry is taken in different forms: cash, furniture, gold ornaments or valuable gifts. Despite dowry being illegal in Bangladesh (GOB 2010), many people follow or are forced to follow this tradition. Poor families try to arrange their daughters' marriage at an early age to save them from paying large amounts of dowry money later. A lack of money to raise a girl at home and insecure poor housing are also important factors. In addition, repeated floods strongly impact chronic poverty and decisions about child marriage, which will be discussed in the next section. For well-off families, a lack of awareness and culture are the main reasons given for arranging girls' marriage at an early age. However, there is a relatively new reason emerging for child marriage, too, as love affairs and elopement are becoming very common among underage young people, with some leaving home to get married at the age of 12 or 13. This behaviour has become a great concern for parents in this village.

The researcher asked a key informant (aged 63 years), who was the Upazila Chairman for a long period and is now working as the Assistant Marriage Registrar, 'is this a new trend in this village?'

TABLE 3.2 Age of marriage in Kashiganj village, Bangladesh

Age (years)	*Below 15*	*15–17*	*18–20*	*21 and above*	*Total*
Female age at marriage	40.1	38.5	18.4	3.0	100.0%
Male age at marriage	0.5%	9.9%	25.3%	64.3%	100.0%

Source: Field survey, 2019

> *Yes, young people getting involved in relationships is a very new thing, we are helpless. If they get involved in any physical relationship, we do not want to, but we have to arrange marriage for them. They may be 13–15 years old...both boy and the girl... what can we do?*
>
> *(Male, Rangpur)*

According to local cultural norms, people cannot be in any physical relationship without marriage. This is a very serious offence. So child marriage is arranged for these young couples. A similar new trend is found among modern youth in Bali where child marriage becomes a solution for young couples wishing to be together (Horii 2020). However, child marriage always has negative effects on economic and social status, on marital relations and on health, particularly for the girl/woman. Child marriage leads to early pregnancy, malnutrition and various reproductive health problems (MacQuarrie and Juan 2019).

Domestic violence

Domestic abuse of wives is another common form of GBV in this village. According to the study, it is very common among poor families, but also present in families from any economic class (see also Ameen 2005 in Ashrafun 2018). Again, dowry is one of the main reasons for physical abuse of wives (Hoque and Kamruzzaman 2016; Nasrin 2011), in order to bring more dowry money from their parents. The respondents report that men who are greedy, irresponsible and do not want to work and earn for the family are more likely to get involved in this form of domestic violence. Some of them are drug or gambling addicts, who raise money by stealing things from their homes or abusing their wives to create pressure for more dowry money.

However, sometimes, there are no underlying factors, as survivors explained during the interviews. Husbands may physically abuse their wives without any evident reason, and this is accepted as normal behaviour by the man himself and by society at large (Anwary 2015). Mashuda explains in her words,

> *My husband started beating me after 2 years of our marriage. He beats me without any strong reason...if I tell him to buy food, he beats me...maybe I was late to fetch his shirt, he beats me...actually he does not need any reason to beat me...I am tired of this.*
>
> *(Female, Rangpur)*

The present study reveals that both men and women are sometimes involved in extra-marital relationships in the village. In particular, men committing adultery or going to brothels are reported to be common. Some women also have extra-marital relationships which are discussed later in this chapter. These behaviours also become a major reason for conflicts between husband and wife, which often lead to physical abuse of women.

Domestic violence is not commonly talked about by survivors with others outside the family. This was a great challenge for the research study. 53.5% of respondents in the household survey mentioned the presence of domestic violence such as the physical abuse of wives. However, according to the key informants and in-depth interviews, it is more common still, and it increases during and after floods.

Sexual abuse, harassment and rape

Eve teasing, cat-calling and sexual harassment are faced by young girls and women in this village. Groups of young boys create insecurity for girls, especially those from poor families who make daily trips to schools or colleges. Young women report feeling unsafe while moving around alone. A case of the rape and killing of a girl was reported shortly after this research was conducted. The body of the girl was found dumped in a wetland near the village. Again, insecurity around GBV is highly interrelated with socio-economic conditions, as the study reveals that young girls from well-off families in the village do not feel this insecurity about their safety, but this is not the case for many others. As we saw in Chapter 2 for cyclone events, the risks of sexual harassment, abuse and abduction increase during flooding for the female members of poor families staying in damaged houses and over-crowded shelters. The impacts of disasters on GBV conditions and women's lives in Kashiganj village are explored in the following sections.

3.6 GBV during flooding and its aftermath

The inhabitants of Kashiganj village experience floods during the rainy and late autumn seasons. Inundation with rain and river water is common every year, however, larger floods occur every two or three years and some of these are very severe. The flood in 2017 was the worst in recent history due to its larger affected area and longer duration. Inhabitants of the village lost crops and cattle, and houses were damaged. These losses were a double burden for the poorer residents alongside their long-term poverty. Many people took shelter at the madrasha, which became a temporary flood shelter. The overcrowded environment of the shelter was very unhealthy and insecure, especially for girls and women. According to 78% of respondents to the survey, the number of GBV cases increased during and after the floods. Women's everyday struggles to protect themselves from GBV become extreme state due to floods. Nasreen's (2010) study also revealed increased rates of violence against women living in three flood-affected regions of Bangladesh. The impacts of floods on the incidence and nature of GBV in Kashiganj village, and the subsequently increased disaster vulnerability, are discussed below, drawing on data collected from the in-depth interviews with GBV survivors.

Women's experiences of GBV in shelters

When flood water increases and enters houses after inundating croplands and roads, people have to leave their homes to take shelter in a safe place. Residents of kutcha houses mostly evacuate, due to the risk of their houses collapsing. When the water level reaches up to their knees, the residents of semi-*pucca* houses also leave their homes. During the 2017 flood, 55.4 % of families left their houses. 35% of families took shelter in the madrasha and others went to relatives' and neighbours' houses. The madrasha is the only shelter in the village and does not have sufficient facilities for a large number of people. There are only two or three working toilets and bathrooms for hundreds of people, and a small cooking space. Four or five families (20–25 people) share each room. The lack of food and the absence of healthcare facilities and medicine increase the suffering of inhabitants of all ages; nevertheless, these problems have the greatest impacts on women.

Sharing a room with men creates huge pressure on girls and women due to a lack of privacy while they are resting, sleeping or breastfeeding. Women reported staring, verbal abuse and unwanted touching from men, making the daily life of a female evacuee difficult. They could not sleep well for long periods. Mothers are always anxious about their teenage girls in case they are harassed or lose their honour (Rashid and Michaud 2000). Young men come from other villages at these times to harass and assault girls. This sometimes also leads to child marriage, as parents become worried about their daughter due to the lack of security. Harassments in the temporary shelters are also mentioned in an IFRC (International Federation of Red Cross and Red Crescent Societies) report on a Namibian flood, where children who were staying in the temporary shelters became vulnerable to GBV in the absence of their parents (IFRC 2015). In another study of flooding, women in Cambodia who took refuge in temporary flood shelters also expressed their concern about being abused by strangers (Chanthy and Samchan 2014).

The lack of sanitation is a problem for everyone at the shelter in Kashiganj, however especially so for women. While men can relieve themselves anywhere and anytime, and this is accepted by society, there are many customs, rules and regulations for women using toilets and taking baths (Rezwana 2018). Whatever emergency is going on, women need a private place. Due to the higher number of users, toilets and bathrooms of the shelters are always busy and dirty. Women have to wait until night to take a bath in an open space or join a long queue to use the facility. This situation is used to abuse women. As Rahima mentioned in her interview,

> *They (men) sometimes intentionally take more time to create pressure on women while using toilets. This is fun for them but humiliation and abuse for a woman.*
>
> *(Female, Rangpur)*

Lack of sanitation becomes more difficult still for expectant mothers, menstruating women and older women. Sometimes, they have to use the floodwater to wash, which results in different types of skin diseases and reproductive tract infections. UN Women's report on flood-affected regions of Bangladesh in 2020 also mentioned the sanitation problems of women staying in the shelters (GiHA 2020).

The lack of cooking space and conflicts over using the facility becomes another struggle for women. They experience constant tension, following a changing routine to cook food for their families. Sharing overcrowded rooms with others, cooking, bathing or going to the toilet at dawn or at night always increases women's risk of GBV. There are some male perpetrators who use this opportunity to abuse women during disaster periods. According to the study, 69% of residents feel that the shelter is not safe for female family members. Most of them do not want to take shelter at the madrasha but have no alternative. However, to avoid these problems, some families stay at their inundated houses or other places that pose other risks, as discussed below.

Women's experiences of GBV in flooded houses

About 44.5% of families stay in their inundated houses during the floods. The reasons for this decision are summarised in Table 3.3.

Among the reasons for staying in inundated houses during floods, avoiding the risk of GBV is the primary one, along with other problems posed by shelters. People are afraid that female relatives will be abused and humiliated while staying at the shelter. The need to look after household things and cattle also prevents some families from leaving their houses during floods. However, staying at damaged and inundated houses is also unsafe – the neighbourhood becomes depopulated as many others leave for shelter, and people have to be alert to burglars and protect girls and young women from being kidnapped.

Women have to collect safe drinking water or relief, often walking a long distance, which also increases their insecurity to GBV (GiHA 2020). They sometimes remain alone at home when men go out for work or to collect relief. Some families plan to send their female members and children to relatives' houses, mostly to their parent's houses, to keep them safe. However, sometimes, women are forced to leave their own house by their husbands to save money and reduce their responsibilities during the difficult time of flooding. This is another form of abuse for some women, as Rina mentioned in her interview,

> *They (husband and mother-in-law) sent me and my children to my father's house during the last flood. I have a stepmom and we were not welcomed. My father married me off at an early age due to poverty. So, staying with him with children for two months become a burden for him. He is too poor, feeding extra four mouths is difficult for him...but I cannot say no to my husband and mother-in-law.*
>
> *(Female, Rangpur)*

TABLE 3.3 Reasons for staying at inundated home during floods

Reason for staying at inundated home during floods		*Percentage (%)*	
Problems in the shelters	Lack of security from GBV for women	38.2	55.8 (total)
	Lack of privacy and over crowding	17.6	
To protect household things		46.6	
More comfortable at home		22.5	
Others		9.4	

Source: Field survey 2019

Rina is a domestic violence survivor who is beaten regularly by her husband. She had to obey her husband and mother-in-law's decision and leave the house during floods, to avoid further abuse. When I went to Kashiganj village to conduct interviews on domestic violence, Rina's situation was mentioned by the neighbours who had heard her screaming.

GBV after flooding: the most crucial time for survival

It is no exaggeration to say that flooding creates a battleground for women in Kashiganj village. After floodwaters have receded, when savings are finished, relief has not been distributed and men cannot earn money, acute poverty strikes as well as violence against women. According to the survivors in this study, poverty has a strong influence on increasing different types of violence: domestic violence, dowry violence, child marriage and trafficking. Nasreen's (2008) research also revealed three main causes of the increasing incidence of violence after flooding to be men's lack of employment (63.5%), the lack of security for women (14.3%) and a lack of food (14.3%).

Domestic and dowry violence

The study reveals that most husband–wife conflicts are worsened by the lack of money during the post-flood period, and these conflicts often end with wife beating. Setara explained the situation she faced during the recent flood in 2017:

> *We could not eat properly during the flood. I have three daughters, one of them became very sick after the flood. My husband did not have any work. We did not have any money and there was no food at home. One day I beat one of my daughters when she asked for food. My husband lost his temper seeing this and started beating me. He took my daughters away from me. I was not allowed to hold my children, although the youngest one needed breastfeeding. She got sick. I had to tolerate everything.*
>
> *(Female, Rangpur)*

According to the survivors, men often lose their main job after flooding, and it takes time to get back to employment. A few are able to arrange temporary work, which is not enough to feed the family and repair the damage to housing and land. Again, some men do not want to change their jobs and some are perceived as being too lazy to search for a new job. When men are confined at home without a job after disasters, often frustrated and anxious, the rate of domestic abuse increases (Chanthy and Samchan 2014; Naznin et al. 2021). However, women need money to buy food for their children. This study reveals that when women talk about buying food and other subsistence at these times, some men become more violent. This appears to be common in domestic violence after flooding. The rate of physical violence against wives increases at an alarming rate, both during and after disasters. In another study after flooding in Benin, Africa, husbands were even seen to be beating their wives in front of other people (CARE 2011). Abusive husbands and women's in-laws may also become more violent to raise dowry money after flooding, resulting in women being beaten when they refuse or are unable to bring more dowry money from their parents.

Habiba was married to a man in another village. Five Lac Taka (5,000 GBP) was fixed as the dowry money, but her father could only pay three Lac (3,000 GBP) during the wedding. Her husband's family started beating her after marriage due to the unpaid money. The violence was so severe that she had a miscarriage and lost her child. Her family had to apply for a divorce to save her from this situation. The severity of wife beating was also mentioned by another victim-survivor, Meena, in her interview; she was beaten severely by her husband and his family who tied her up and gagged her so that no one could hear. On one occasion, Meena reports her husband setting fire to her during a beating.

The involvement of other family members alongside husbands in violence against women can be explained by these members (such as mothers-in-law) often lacking resources, and themselves experiencing less power in the family, stress and frustration. 'These family members are likely to commit violence against members with more resources and power to compensate for their lack of power' (Lenton 1995 in Anwary 2015: 39).

Child marriage

Child marriage, as a form of violence against girls, also increases after flooding at an alarming rate. Many parents marry off their young daughters after flooding to save on daily expenditure and reduce the work involved in protecting them from GBV and losing their honour. Sometimes, they are so desperate that they even take their daughters to relatives' houses far from their village, in order to hide the girl's age and arrange a marriage. Lesser dowry money is asked for young girls, and as explained above, parents want to take this opportunity. However, some men use this crisis period created by flooding intentionally to marry young girls. Even cases of middle-aged men marrying girls of 13–14 years of age were reported in this research.

Marriage at an early age is also sometimes arranged in order to try to keep girls safe. A grieving mother of a young girl who had recently died, explained the situation:

> *I could not keep my daughters at home, I had to marry them off at a very early age, at 13–14 years. They were very pretty; men were after them and disturbed them a lot. I felt insecure. One of my daughters was living with her husband in other village. She had a two-year old son. One day she committed suicide, we do not know the reason. We also do not know if this is a suicide or murder. We did not get any suicide note. We knew she was happy with her life but she left us for forever.*
>
> *(Female, Rangpur)*

BOX 3.1 CHILD MARRIAGE, DOWRY VIOLENCE AND ATTEMPTED SUICIDE

Hasina was married off at the early age of 12–13 years when she was in class seven at school. Her father had to pay dowry money during the wedding. However, Hasina's in-laws started torturing her for more money after the marriage. She was beaten severely. Her parents had to send more money several times to save their daughter from violence. However, her husband was not happy with the amount of money and abandoned her. After the divorce, she returned to her parent's house but she was not welcomed. She became an extra mouth to feed. Her family humiliated and neglected her. Hasina attempted to commit suicide but luckily was saved on that occasion.

Child marriage ruins girls' lives and increases their vulnerability to domestic violence (Ameen 2005 in Ashrafun 2018; and see the case of Hasina, Box 3.1); they cannot complete school education and instead become young mothers and dependent on their husbands. One victim of child marriage, Mashuda, described her situation:

> *For my fate, I blame my father...he married me off at the age of 13, I could not finish my education, he had to pay 30,000 Taka (30 GBP) as dowry...my husband beats me a lot, one day he struck me so hard that I needed medicine to recover...my parents-in-law are very good...they try to stop their son but failed...I cannot change my condition as I could not finish my education, I am not allowed to go outside to earn money as a day labourer even though we are very poor... my husband is very lazy... does not want to earn money.*
>
> *(Female, Rangpur)*

Forced prostitution

Forced prostitution is another type of GBV which increases after floods. Women from poor economic backgrounds are sometimes forced by their husbands to get involved in sex work. It was reported by respondents that husbands who are drug addicts, gamblers and/or do not earn enough are more likely to depend on their wife's earnings from this profession. Residents of Kashiganj village know about this form of violence against women but most do not talk openly.

Migration and trafficking

Youth also become responsible for earning money after disasters (Tobin-Gurley et al. 2016). Boys and girls aged 11–14 years sometimes leave their homes after flooding and migrate to cities to earn money. Boys work in car garages, brickfields or construction projects, and girls work as housemaids. Some boys had accidents while performing this risky work. Girls face high risks of being trafficked or experiencing violence such as rape and sexual assault after migrating (Enarson et al. 2018). Mostly, girls are sent to the cities to work via a middleman, without any proper contract. The study reveals that both ignorance and negligence lead parents to take this risky decision about their children. Similar practices are found in other South Asian countries such as India and Nepal, where false job promises are made to traffic young girls and women from disaster-hit regions (Dutta 2017).

Flooding and male abuse survivors

Men also sometimes suffer domestic abuse, and although the numbers appear to be very low and it tends not to be as serious as for women, it is important to explore it. In the study, the verbal abuse and humiliation of men by their wives, especially after wives' extramarital affairs, were reported. Some male respondents perceive both husband and wife to be responsible for conflicts, suggesting that they both become bad-tempered and tense due to increased poverty during floods, which triggers arguments and violence.

In fact, there are very different perceptions present in the study among the male respondents when it comes to this abuse against men; however, all of them mentioned wives' contribution and fault in triggering domestic violence. According to a male respondent,

> *Men also become victims of domestic violence. You cannot tell everything. A wife was shouting at her husband that he should leave her, so that she could go to her parent's house. The wife has a job while the man lost the job. This is violence against men.*
>
> *(Male, Rangpur)*

This quote represents men's perceptions of abuse against them, with the wife's verbal abuse being named as significant violence by this male respondent when he was asked in the survey about GBV against men in Kashiganj village.

3.7 Gendered power relations and GBV

> *He always behaves very badly with his wife. Tells her off with ill words. He beats her without any good reason. Sometime it becomes severe. However, he never feels sorry for his behaviour.*
>
> *(Female, Rangpur)*

Returning to the more common problem of violence against women, the quote above describes a commonly reported situation as regards male perpetrators in the Kashiganj village. Many respondents report that men who are used to beating their wives never feel sorry or guilty; on the contrary, their behaviour is often accepted by the family and wider society. Gendered behaviour, roles and responsibilities are quite complex in this village and influenced by prevailing socio-economic and disaster conditions. Unlike in some communities, women may work outside the home, they can visit the bazaar, healthcare centres, towns and friends' houses without a chaperone. Their freedoms therefore differ from those of women in Barguna (Chapter 2). Nonetheless, there is always an invisible boundary for women, created by the family or wider society, which raises questions about women's actual empowerment status. The following section elaborates on this point.

Masculinity, women's empowerment and cultural aspects of gender-based violence

The present study reveals a complex situation in Kashiganj village, with a tension between women's freedom and social expectations that leads to different forms of GBV within this patriarchal society. An important example was identified during the fieldwork, and is described below.

Kashiganj is a village of mixed culture; both urban and rural cultural elements can be seen here. As described earlier in the chapter, the use of technology such as mobile phones, the internet, television and international channels is common in the village. Both conservative and semi-secular attitudes (af Buren 2015) are found among the residents. Women are found to be working in the fields, taking cattle to graze and going to the bazaar and nearby cities without any chaperone, as described above. They do not feel hesitant or any cultural barriers in talking with strangers. Some women wear Burka and hijab but many of them simply cover their hair with saree or do not cover. Some women follow new styles in dress and makeup. Women hang around with their female friends and relatives in the afternoon (Figure 3.2) and spend time together walking in the walkways and enjoying street snacks from the temporary food stalls in the village. However, there are some families, very few in number, who do not allow their female members to follow these new trends and changing culture. These women remain at home most of the time. The study reveals mixed opinions about these new trends in the village – some residents are tolerant of them, some view them as a bad thing, while many wonder whether these new trends are the main reason for increasing numbers of extra-marital relationships and love affairs among young people.

Jeba is one of the young married women who likes to dress in the new style and hang around with friends. She did not complete her school education. She was living with her husband and two children near her parents' house. Her first husband was a middle-income businessman. Her style and choice of friendship circle (both male and female) had created a mixed reputation for her among the villagers: she is called trendy, spoilt and a woman who is a bad influence. One day she had a huge fight with her husband because of this. It is reported by other respondents that she reprimanded her husband who, the next day, committed suicide. A village meeting was called in response to her mother-in-law's appeal, in which Jeba was accused and given a punishment. She was tied to a tree and beaten severely in front of everyone. Later, a video of this punishment went viral and a case was filed against male members of this village. After that day, police protection was arranged for Jeba to protect her from further violence. As many of the male members of this village are among

Women during leisure time

FIGURE 3.2 Leisure time and friends

the accused in this case, it was difficult to get information during the fieldwork. Most of the villagers did not want to talk much on this topic. Jeba later remarried and is now living with her second husband.

Jeba's case reveals some important points to consider. It raises questions about the status of women's empowerment and the freedom they enjoy in choosing dress, daily movement and expressing their opinions. Jeba was taken and judged, accused and punished without any involvement of the legal authorities. The reason for her husband's suicide might be that she abused him, or he may have felt humiliated, or some unrelated personal situation: this is not identified clearly as the case is still in court. However, Jeba was accused and beaten by the village committee without any further thought.

Here, it should be mentioned that there were other cases of casualties in this village: a young woman committed suicide in her husband's house and a girl attempted to commit suicide in her parents' house. Both of them were married and known to be victims of domestic violence. There was no meeting arranged by the village committee to judge their parents or husbands, nor has any report been made to the legal authorities. Besides these cases, as reported earlier, wife beating is very common in this village and survivors rarely get any attention from wider society. So, these cases show the gendered social attitudes towards those suffering GBV; male survivors get rapid attention and support, whereas female survivors are rarely considered worthy of attention and receive very little help. From the outside, women in this village seem to be enjoying rather more freedom than many others in some parts of Bangladesh, but the question arises – Are they really empowered? How much freedom a woman and man can actually enjoy is determined by social norms, gender disparities and patterns of discrimination. Men are not punished for visiting brothels or having extra-marital affairs, whereas women are often accused and beaten if their husbands become suspicious about their actions or character. Society is frequently judgemental about women, their choices, attitudes and behaviour. Even the way women dress, their style and way of talking can lead to a bad reputation. This is a dilemma for women in this village, and it is always a risky decision they make as to how much freedom they should enjoy. According to Anwary (2015), when structurally powerless women upset and challenge normative conceptions of femininity and masculinity, and create symbolic space for alternative identities, 'their action may not be empowering for themselves' (Anwary 2015: 44); instead, it may increase their vulnerability to GBV, as this study also demonstrates.

Male approval as a precondition of women's empowerment, in relation to GBV

Women's age is related to their empowerment in rural societies of Bangladesh. In most cases, older women gain decision-making powers in the family. Especially after her son's marriage, household dynamics change and the mother-in-law achieves a higher position. However, this position and power are sometimes used against daughters-in-law through involvement in domestic violence. As this study has shown, husbands along with their families (mother, father or sister) may get involved in dowry violence. Women, in this case, get involved in violence to ensure their own survival and security within a social context that is shaped by men (MAHR 2003). However, woman's practice of power in the family first needs the approval of the head of the family, who is usually a man. This is the precondition of a woman's empowerment. For example, if there are any problems between a daughter-in-law and her husband's parents, the husband may take the decision either to live separately with his wife or to keep his wife in that risky environment. However, a widow with an abusive daughter-in-law has to continue to live with her, if her son does not permit her to take charge, or must live separately (Powell et al. 2009). The study in Kashiganj village has revealed some cases where aged mothers-in-law are abused by their daughters-in-law, and this violence also increases during and after floods. Therefore, age does not always dictate women's empowerment; rather, approval from male members of the family is the precondition of women's power and position in the family, which again shows the predominant influence of masculinity in a patriarchal society.

Social status, power and GBV

Social status, which may include education, income, family wealth and reputation, has a strong influence on the vulnerability of women to GBV. This study reveals (according to KII and FGDs) that education and income status help women to be empowered and keep them safe from domestic violence like wife beating. Besides, young girls and women from elite families do not feel vulnerable while travelling to an educational institution, or to other areas. They can easily move around in the daylight without any fear of eve teasing or sexual assault, whereas most of the women of this village do not have this privilege. One of the young women explained in the interview, '*They [abusers] do not have the courage to disturb me at all; they all know me and my family. I never felt any problem*' (Female, Rangpur). This is also true for women who are employed in jobs such as school or college teacher, or who are relatives of an elite family. Local gangs do not have the courage to fight against their family members. This situation clearly shows the complex relation of power, politics and GBV. While GBV is the malpractice of power by men against women, women's education, income and family status can help to reduce their vulnerability to violence.

Poverty, flooding and GBV – a vicious cycle

The discussion above shows the power of wealth in reducing women's vulnerability to some types of GBV, even in a highly patriarchal society, whereas poverty does the opposite. The female members of poor families are the most vulnerable to GBV in this village. Child marriage, incomplete education, lower income or no income increases the vulnerability of women to domestic violence, which in turn increases in times of poverty and disasters. Girls from poor families are mostly married to men in poor economic situations, and regular floods from the Jamuneshori River are a strong factor in the persistent poverty in this village. This contributes to keeping many women trapped in a vicious cycle of poverty, flooding and GBV. It becomes very difficult for survivors to escape this cycle and improve their lives, without proper help and support.

3.8 Conclusion

As we saw in relation to cyclones in Barguna in Chapter 2, women's greater vulnerability to flood disasters and to GBV compounds the cyclical relationships between them. In this chapter, these relationships have been explored from a study of the flood-prone village of Kashiganj in the Rangpur district. Poor economic conditions, weak infrastructure, and the lack of shelter facilities increase flood-related problems in the village. However, disparities in education, employment and income, along with high incidence of different forms of GBV during non-disaster periods, increase women's vulnerability to both floods and the poverty that can follow. Young girls are forcibly married or sent away from home to generate income;

women are beaten for dowry money or forced into prostitution; young women and girls face insecurity from rape and harassment in shelters and flood-damaged homes. All these intensify the problems that floods bring for women. In many cases, these incidents remain hidden within the family and are not reported to authorities (this is discussed in more detail in Chapter 7). Kashiganj women are allowed to travel alone, may work in the fields and visit friend's houses, but their freedoms are still conditioned and influenced by patriarchal society. Their apparently greater empowerment in comparison with, for example, Barguna women (Chapter 2) does not help to protect them from high rates of violence that are part of the vicious cycle of poverty, GBV and flooding. In Chapter 7, we discuss the action and support required from the family, society, and local and national authorities to reduce their vulnerability to both GBV and floods.

References

af Buren, A (2015). "Living Simultaneity: on religion among semi-secular Swedes." Dissertations published by the Department of Literature, History of Ideas, and Religion, University of Gothenburg, 41: 1102–9773.

Ameen, N. (2005). *Wife Abuse in Bangladesh – An Unrecognised Offence*. Dhaka, The University Press Limited.

Anwary, A. (2015). "Construction of hegemonic masculinity: violence against wives in Bangladesh." *Women's Studies International Forum* 50: 37–46.

Ashrafun, L. (2018). *Women and Domestic Violence in Bangladesh: Seeking a Way Out of the Cage*. London, Routledge.

Azad, A. K., Hossain, K. M., & Nasreen, M. (2013). "Flood-induced vulnerabilities and problems encountered by women in northern Bangladesh." *International Journal of Disaster Risk Science* 4(4): 190–199.

Banglapedia (2015). Flood. *Banglapedia: National Encyclopaedia of Bangladesh*. http://en.banglapedia.org/, accessed on 28/07/2020.

BBS (2013). *Population and Housing Census 2011, Community Report, Zila: Rangpur*. Bangladesh, Bangladesh Bureau of Statistics.

BBS (2016). *Report on Violence Against Women (VAW) Survey 2015*. Bangladesh, Bangladesh Bureau of Statistics.

CARE (2011). An assessment of gender-based violence in emergencies in southern Benin, Emergency Program, CARE Benin.

CARE (2020). Feed the Future Ethiopia, Resilience in Pastoral Areas (RiPA North), Rapid Gender Analysis for Afar Flood, USAID, Mercy Corps and CARE.

Chanthy, S. & Samchan, H. (2014). *Flood impacts on women: exploring the possibility of gender sensitive DRR planning*. Actionaid, Cambodia.

DAWN (2021). "Domestic violence against women, children increased during lockdown: report." https://www.dawn.com/news/1611642/domestic-violence-against-women-children-increased-during-lockdown-report, accessed on 06/06/2021.

Dutta, P. M. (2017). "Natural disaster and vulnerability to trafficking of women and girls in India." *European Scientific Journal* 7881: 54–69.

DWAG (2020). "Ongoing floods expose women to further risk of gender-based violence." http://www.darfurwomenaction.org/press_release/ongoing-floods-expose-women-to-further-risk-of-gender-based-violence/, accessed on 06/06/2021.

Eckstein, D., Hutfils, M. L. & Wings, M. (2018). *Global Climate Risk Index 2019*. Berlin, Germanwatch. https://germanwatch.org/en/16046, accessed on 05/06/2021.
Enarson, E., Fothergill, A., & Peek, L. (2018). "Gender and disaster: foundations and new directions for research and practice." In Rodriguez, H., Donner, W., & Trainor, J. E. (eds.) *Handbook of Disaster Research*. Cham, Springer: 205–223.
GiHA (2020). Preliminary rapid gender analysis of monsoon flood 2020, Gender in Humanitarian Action Working Group, Bangladesh, UN Women.
GOB (2010). The Domestic Violence Bill-2010. Ministry of Women and Children Affairs, Bangladesh, Government of Bangladesh.
Hoque, M. M. & Kamruzzaman, M. (2016). *From kitchen utensil to sewing machine: women's internal migration and gender dynamics in rural Bangladesh*. GIGD, BRAC University, Bangladesh.
Horii, H. (2020). "Child marriage as a 'solution' to modern youth in Bali." *Progress in Development Studies* 20(4): 282–295.
IFRC (2015). Unseen, unheard: gender-based violence in disasters, global study. *Researchers*: Dzieszkowska, A., Graven Minnig, S., Lee, A., Mangwende, F., Manning, L., Nicol, Ch., Stern, S., Trapman, I. International Federation of Red Cross and Red Crescent Societies, Geneva, Switzerland, www.ifrc.org.
Lenton, R. L. (1995). "Power versus feminist theories of wife abuse." *Canadian Journal of Criminology* 37: 305–330.
MacQuarrie, K. L. & Juan, C. (2019). "Trends and factors associated with child marriage in four Asian countries." *Gates Open Research* 3(1467). https://gatesopenresearch.org/articles/3-1467, accessed on 30/03/2022.
MAHR (2003). "What is gender-based violence? Minnesota advocates for human rights." http://www.stopvaw.org, accessed on 1/10/2020.
Mason, J. (2002). *Qualitative Researching*. London, Sage Publications Limited.
Memon, F. S. (2020). "Climate change and violence against women: study of a flood-affected population in the rural area of Sindh, Pakistan." *Pakistan Journal of Women's Studies: Alam-E-Niswan* 27(1): 65–85.
Nasreen, M. (2008). "Violence against women during flood and post-flood situation in Bangladesh." Bangladesh, ActionAid.
Nasreen, M. (2010). "Rethinking disaster management: violence against women during floods in Bangladesh." In Dasgupta, S., Iriner, I., & De, P. S. (eds.) *Women's Encounter with Disaster*. London, Frontpage: 232–244.
Nasrin, S. (2011). "Crime or custom? Motivations behind dowry practice in rural Bangladesh." *Indian Journal of Gender Studies* 18(1): 27–50.
Naznin, Z., Sharmin, M., & Mohan, F. (2021). Addressing gender-based violence in Bangladesh, Flood Resilience Portal. https://www.floodresilience.net/blogs/addressing-gender-based-violence-in-bangladesh/, accessed on 01/06/2021.
NIRAPAD (2017). "Bangladesh: Flood Situation-August 22, 2017." www.reliefweb.int, accessed on 29/07/2020.
Philip, S., Sparrow, S., Kew, S. F., van der Wiel, K., Wanders, N., Singh, R., Hassan, A., Mohammed, K., Javid, H., Haustein, K., Otto, F. E. L., Hirpa, F., Rimi, R. H., Saiful Islam, A. K. M., Wallom, D. C. H., & van Oldenborgh, G. J. (2019) "Attributing the 2017 Bangladesh floods from meteorological and hydrological perspectives." *Hydrology and Earth System Sciences* 23: 1409–1429. doi:10.5194/hess-23-1409-2019
Powell, S., Plouffe, L., & Gorr, P. (2009). "When ageing and disasters collide: lessons from 16 international case studies." *Radiation Protection Dosimetry* 134(3–4): 202–206.

Rahman, S. (2019). "1,787 cattle, poultry farms affected by flood: report." *The Financial Express*. www.financialexpress.com.bd, accessed on 23/08/2020.

Rashid, S. F. & Michaud, S. (2000). "Female adolescents and their sexuality: notions of honour, shame, purity and pollution during the floods." *Disasters* 24(1): 54–70.

Rezwana, N. (2018). *Disasters, Gender and Access to Healthcare: Women in Coastal Bangladesh*. Oxon, UK, Routledge, Taylor & Francis Groups.

Robson, C. & McCartan, K. (2016). *Real World Research*. Chichester, UK, John Wiley & Sons.

Schlein, L. (2018). "Study: sexual, gender-based violence rises after natural disasters." https://www.voanews.com/east-asia/study-sexual-gender-based-violence-rises-after-natural-disasters, accessed on 06/06/2021.

The World Bank (2020). "COVID-19 to add as many as 150 million extreme poor by 2021." https://www.worldbank.org/en/news/press-release/2020/10/07/covid-19-to-add-as-many-as-150-million-extreme-poor-by-2021, accessed on 28/05/2021.

Tobin-Gurley, J., Cox, R., Peek, L., Pybus, K., Maslenitsyn, D., & Heykoop, C. (2016). "Youth creating disaster recovery and resilience in Canada and the United States: dimensions of the male youth experience." In Enarson, E. & Pease, B. (eds.) *Men, Masculinities, and Disaster*. New York, Routledge: 152–161.

UNICEF (2018). *Global Databases: Child Marriage Data: Percentage of Women Aged 20 to 24 Years Who Were First Married or in Union before Ages 15 and 18*. New York, UNICEF. https://data.unicef.org/topic/child-protection/child-marriage/, accessed on 07/08/2021.

4

FLOODING, MENTAL HEALTH AND CRIME

Women's experiences in England

Rachel Pain and Nahid Rezwana

4.1 Introduction

The last two chapters focused on women's experiences of extreme weather events leading to disasters and raised levels of gender-based violence in Bangladesh, a low-income country. As a growing literature has documented, this relationship between disasters and GBV is also common in medium- and high-income countries (see Chapter 1). In every part of the world, when disaster strikes, women are among those groups who are particularly at risk, both from the immediate impacts of the disaster and from its longer-term consequences. Heightened levels of GBV are almost always part of this layered disaster.

In this chapter, we report on original research conducted in two settlements in Cumbria in North West England which were badly affected by flooding in December 2015. In the UK, more widely, both disasters and gender-based violence are popularly assumed to be less problematic than in other parts of the world. Nonetheless, as we stated in Chapter 1, GBV is a truly global issue, with high rates in Europe and North America (World Health Organization 2021). In England and Wales, in March 2020, 1.6 million women were victims of domestic abuse (ONS 2020a), and an estimated 4.9 million women had been victims of sexual assault at some point in their lives (ONS 2020b). As we discussed in Chapter 1, previous studies suggest that rates of GBV also rise when disasters take place in higher-income regions of the world, such as during Hurricane Katrina in the USA (Schumacher et al. 2010), after earthquakes in Japan and New Zealand (Masson et al. 2016; Phillips 2011).

The chapter focuses particularly on the consequences of the floods on women's mental health, showing that flooding and its impacts create both short and long-term trauma. It also reports on crime experienced by women during and immediately after flooding, including some cases of GBV, as well as impacts on

DOI: 10.4324/9781003089780-4

their finances, physical health and caring responsibilities. We show how one trauma compounds another, in ways that amplify the layering of disasters that we described in Chapter 1. The research also offers some important insights on women's perceptions of the link between flooding and domestic abuse, shedding light on general understandings of GBV among these communities.

4.2 Gender-based violence and disasters in the UK context

Because of its geographical location, the UK has experienced far fewer disasters that are partly caused by natural forces than highly disaster-prone countries such as Bangladesh. While serious and widespread disasters of any type have been relatively rare in the UK (with the exception of the Covid-19 pandemic; see Chapter 5), the disasters from which it is most at risk are heatwaves and flooding, both of which are becoming more frequent due to climate change.

Before the Covid-19 pandemic, there was very little research on the relationship between disasters and gender-based violence in the UK. As in many countries, understanding the connection has been made more complex by the lack of consistent recording practices (see Parkinson et al. 2011). As Akerkar and Fordham (2017) argue, it should not be assumed that in countries such as the UK, which are often assumed to be more gender equal, women and men experience similar disaster impacts. Moreover, since the Covid-19 pandemic began in 2020, it has become even clearer that the UK is not even close to being a gender-equal society, and indeed that in many areas of life, women's struggles for equality are slipping backwards (Johnston 2021). The pandemic has negatively affected many areas of women's lives, more so than for men (see Chapter 5). This is especially the case for those with caring responsibilities, black and racialised women, those with disabilities and LGBT women (Womens Policy Group Northern Ireland 2020). Prior to Covid-19, however, there was relatively little research on women and disasters in the UK, particularly on trauma or GBV. Akerkar and Fordham's (2017) research on floods in two counties of the UK in 2007 and 2008 showed that men and women both experienced flood impacts, such as mental health effects, but that these, and their paths to recovery, were different. We compare our findings to this previous study throughout the chapter.

4.3 Study area

Cumbria is in the far North West of England, one of its largest counties but with the second-lowest population density. It is predominantly rural, with few urban areas, and famous mountain and lake scenery that make it very popular with tourists. It was selected for the study because, as a large rural area, it is more difficult for GBV survivors to get support, report and escape abuse (Little 2017); but also because it has suffered severe flooding in recent years. Two settlements that were most affected by the serious floods in 2015 were selected for study: the small city of Carlisle in the north of the county, which had a population of 108,200 in 2015

(Cumbria Intelligence Observatory 2015), and the town of Kendal in the south, which had a population of 36,731 (Cumbria Intelligence Observatory 2017). The research was conducted over the summer of 2019, asking respondents to reflect back on the flooding and its impacts.

The floods in question occurred on 5 and 6 December 2015, as a result of Storm Desmond which affected Cumbria and other parts of Northern England. Unprecedented prolonged, intense rainfall fell onto ground that was already saturated after a very wet autumn. The UK's 24-hour and 48-hour rainfall records were both broken in Cumbria during the storm. Carlisle is located on the confluence of three rivers, while Kendal is immediately downstream of another confluence of three rivers; both towns have long histories of flooding. The 2015 floods were the greatest ever seen due to climate change (Otto et al. 2018), and the Rivers Eden (which runs through Carlisle) and Kent (in Kendal) reached their highest ever recorded levels. In both towns, the existing flood defences were overtopped, although the defences did slow the pace of flooding which gave some households more time to prepare. Approximately 2,100 properties in Carlisle were flooded (Environment Agency 2016a), and 2,150 in Kendal (Environment Agency 2016b).

Cumbria also experiences high rates of domestic abuse. It is estimated that over 15,000 women and girls in Cumbria experienced domestic abuse over the year 2018–2019, with the majority of child protection cases in the county involving domestic abuse (Cumbria County Council 2019). Figures have risen rapidly in recent years (Whitehaven News 2019), although this may be likely to reflect a greater willingness to report abuse. On average, Cumbria Police deal with over 17 incidents of domestic abuse per day, with around half of these known to be repeat victimisation (Cumbria Police 2019), and domestic abuse is thought to cost £48 million in direct costs (Cumbria County Council 2019). It is important to highlight the fact that most cases of domestic abuse are never reported to police in the UK (HMIC 2015), so these statistics are likely to be well below the actual incidence. Moreover, in a rural region with fewer services than in larger UK cities, it is more difficult for victim-survivors to report abuse and get support.

Our research aimed to explore the experiences and perceptions of flooding, domestic abuse and other crimes among female residents who were affected by the floods in 2015.

4.4 Methodology

The study presented particular methodological challenges for the research team. In contrast to the field studies that had been successfully completed in Bangladesh using face-to-face surveys (Chapters 2 and 3), this method was considered less likely to elicit frank open responses on this topic in the UK. Instead, female respondents were accessed by postal survey in Carlisle and Kendal. This method was chosen to allow privacy in answering, although it also has some limitations which are discussed below. As four years had passed, it was important to target as many of those who had been affected by flooding in 2015 as possible. Using Environment

Agency flood maps for the two areas which record the boundaries of the maximum levels of the floods, all the streets and hamlets within the boundaries of Carlisle and Kendal that were affected by flooding were identified. We then mailed a survey to all addresses in these streets, with instructions that only females over the age of 18 who had lived in the same area in December 2015 should complete it, enclosing a prepaid return envelope. The survey contained a mix of closed and open questions on four themes connected to the flooding: housing issues, economic issues, health issues and crime/assault. Many respondents wrote answers to the open questions at length, creating a large qualitative dataset to supplement the quantitative data. Only some of the data are drawn on here – in this chapter, we especially focus on issues of mental health and crime/assault.

In total, 1,685 surveys were sent out and 380 were returned, giving a response rate of 22.5%. This was reasonably successful given that some residents would have moved away from their homes since the floods four years ago, whether by choice or necessity. Of those who responded, three women reported that they had experienced domestic abuse *during the 2015 floods*: an incidence rate of 0.8%. While this incidence may seem low, the survey did not aim to provide a representative picture of the *prevalence* of domestic abuse, given that the question was very specific about the timeframe in question. Moreover, those women who experienced domestic abuse are also more likely to have moved house since, and therefore less likely to be captured by the survey. The fact that those responding to the survey were over-represented by homeowners (91.9% compared to 8.1% who rented their home), women aged over 45 (77.9%) and those whose occupation is retired (39.7%) may also suppress the reporting of domestic abuse. Further, the survey method may have deterred respondents from reporting very intimate experiences of GBV, although other crimes and sexual harassment were reported.

The data cannot support inferences about the prevalence of domestic abuse during flooding, then. However, they do provide valuable quantitative and qualitative descriptions of, firstly, the mental health effects of flooding, including trauma, and secondly, women's perceptions of the link between domestic abuse and flooding. The data lead us to our main argument in the chapter, which is that when domestic abuse does happen during flooding events in the UK, the impacts of flooding and its mental health effects are likely to exacerbate domestic abuse and trauma and worsen their effects. As concluded from the research in Bangladesh, domestic abuse and flooding are locked in a cyclical relationship, with implications for emergency responses (see Chapter 7).

4.5 The impacts of flooding on housing and financial status

Flooding commonly produces and exacerbates a range of social and economic problems for those households affected. For many of the respondents to the survey, the experience of having their home flooded impacted their housing status, their finances, their health, and their vulnerability to crime. While causing material hardship, these factors also compound each other and have significant impacts on mental health, contributing to the trauma that many studies have shown is associated with experiencing a disaster such as flooding.

In terms of housing, almost half of the 380 respondents said they had had to move out of their homes because of the flooding (46.6%). Of these, over half moved into a friend's or relative's home (54.8%), 17.5% moved into a hotel or Bed and Breakfast accommodation, and only 2 (0.5%) moved to an emergency shelter. While around a fifth (21.5%) stayed in temporary accommodation for a period of a few days to a few weeks, the majority (79.6%) were unable to return to their homes for a few months or longer.

> *So many houses were flooded (for the second time in my case) that it took some time to find suitable rented accommodation.** *We lived in caravan on our drive.*
>
> *First day we stayed in a hotel, but could only afford one night, so moved in with my granddad in a two-bed flat. Sleeping on camp beds.*
>
> *We went back home once the water level had dropped. We lived upstairs throughout the refurb [refurbishment].*
>
> *We wanted to move back into our home, even though we stayed in one of the bedrooms, no floors downstairs and far from finished. We need to see what was happening to our home.*
>
> *Unlike our neighbours, we only stayed out a few days. We then moved back in and got to work on the house – we lived upstairs for a number of weeks and ate out a lot.*
>
> *After five months our home was still not fully repaired, but we moved back in to get back to normal routines and life.*
>
> *All respondents were female and from Cumbria.

Perhaps surprisingly – and reflecting the fact that the sample is older and has more stable housing tenure than residents of the two towns in general – only 18.4% of respondents said that the flooding caused problems with money, or made these problems worse. Because of the age and tenure profile of the respondents, most homes were covered by insurance policies, and in addition, grants were available for some renovations:

> *We were well covered by our insurance, who paid the rent on our temporary accommodation and the local council were also supportive. Also supported by the Cumbria Community Foundation.*
>
> *Insurance and flood relief were very supportive on boiler costs and surveys. There was council tax relief for 3 months in January.*

However, for those who were affected financially, there had been significant costs involved that had quite severe long-term effects:

> *Paying mortgage and rent for 18 months had a huge financial impact.*
>
> *Insurance pulled out due to the clause 'covered flood damage but not flood water damage.' Had to get a loan to rebuild our home.*

Very upsetting – had insurance but they wouldn't pay for the flood damage.

I lost my business because of the floods.

I gave up work to deal with the flood – there are issues that are still not resolved three and a half years later.

My husband had to retire early due to stress caused by flood insurance and house reinstatement.

4.6 The impacts of flooding on physical health, caring responsibilities and mental health

In terms of health, 16.1% of respondents reported having existing health issues at the time of the flooding, and 5% said they had a pre-existing disability. Almost a third (32.9%) of all respondents said that the flooding had made their physical or mental health worse.

In the open comments, the physical health conditions that were mentioned most often as having worsened were arthritis, heart conditions and diseases of the respiratory system, which predominantly affected older people. Some said their physical health had been worsened by the physical situation they were in after flooding, for example, from living in a house that was damp or had no heating (the flooding took place in winter), or from moving heavy furniture upstairs. Many others noted the relation between physical health and the increased psychological stress associated with flooding:

I had asthma – dampness made it worse.

Post-flood, I was diagnosed with Angina and 10 months later suffered a heart attack.

Being stressed with the day to day events made my rheumatoid arthritis flare up badly.

Asthma due to increased anxiety, and resistance to other illness decreased.

I have ME/Chronic Fatigue Syndrome and it was made much worse by the stress of the flood.

[I have] Chronic Obstructive Pulmonary Disease and the stress accentuated it.

I had a stroke.

I lost a lot of weight and I was so stressed out I ended up having to have a pacemaker operation [for heart disease] while I was in temporary accommodation.

I have a heart condition, stress of dealing with flooding and insurance contributed.

Experience of flooding definitely has a health impact (from speaking to family/friends – they've found the same) – it's a huge shock to the system, physically, emotionally etc.

There were also certain health conditions mentioned that were unique to women. Fourteen women who responded to the survey had been pregnant or looking after a baby when the floods hit, and several say that this added to the mental health effects of flooding:

The stress of dealing with house, insurance company, builders while pregnant.

I was 38 weeks pregnant with my first child. Luckily, I had a healthy pregnancy and I was well and active. My blood pressure increased which was expected.

More stress with a new baby only six months old.

Moreover, in the UK, women are far more likely than men to be primary carers for others, whether for children, parents or their partners. The flooding increased these responsibilities and, again, led to elevated psychological stress as a result:

My partner has health issues and a disability which impacted me as his carer.

This has impacted on my ability to care for my husband without our usual aids, etc, in our own flat.

Stress looking after my partner who is partially sighted.

It affected the mental health of my 12 year old son who suffered from depression.

[I had] stress and anxiety, but I didn't see my GP. There was no point, I was too busy sorting everyone else.

Indeed, in the open comments on health issues, mental health effects were mentioned more often than physical health. Of these, stress, anxiety and depression were the most frequently described, by 22.9% of respondents. For some, this was associated with the immediate impacts of the flooding and its consequences:

It was extremely stressful. There was heightened anxiety about how soon the house would dry out to start repairs. There was shock at experiencing flood water coming in.

Stress of dealing with power cuts/loss of property, plus supporting affected family, friends and neighbours.

As I was 88 years old and living alone, the stress of the flood was very traumatic. I was unable to visit my home for six months.

Many reported sadness at the damage to their homes and precious possessions:

The loss of personal things is heartbreaking – down to the photos of the family that no one else had copies of.

I lost a lot of sentimental items which can never be replaced.

Loss of over 37 years of memories – thrown out with the flood water.

I found it very hard to put belongings out on the street for all to see. Local radio presenter commented it was awful to see streets full of rubbish. How insensitive; this rubbish was my home.

Many others reported that the mental health impacts of the floods had been long-lasting:

We managed the 2005 floods really well. However 2015 floods had a huge impact on my emotional well being, suffering from stress and anxiety.

I now suffer from depression and anxiety.

I still suffer from anxiety and depression which I did not have prior to the floods.

It severely affected my mental health and I had to have 3 months off work due to stress and anxiety.

It became harder and harder to cope over 2016 and subsequently, even after we came home.

A few respondents reported more serious mental health impacts:

Schizophrenia – episode triggered by flood.

Severe depression and a suicide attempt.

My health was very good prior to the flooding. I became very depressed and stressed following the flooding of my home and business...I became suicidal.

For many women, their descriptions of their long-term mental health in the four years since the floods include symptoms of trauma that have not gone away. While some say they have been diagnosed with PTSD (post-traumatic stress disorder) because of the disaster, others do not name this but still recount symptoms such as panic attacks and fear which, when experienced alongside anxiety and depression, signal trauma:

Still suffering from stress three years on. Panic when it rains, will it happen again? – as I could not go through this again, still taking anti-depressants.

I began to experience severe panic attacks and had to take 8 weeks off work. I experience PTSD now whenever it rains for long periods of time.

Stress and horror – long term PTSD. Every time it rains we check out river levels. All our neighbours feel the same way.

Every time it rains for a long time I worry it's going to flood again, so I'm on the edge until the water drains away. It's an awful feeling.

If it rains for a few days in a row I start to panic.

Stress, anxiety, weight loss, insomnia, panic, worry every time it rains!

It worries me that all I had worked for over the years would be lost, and I find that I panic now if it rains hard for a long time. I have panic attacks for no reason.

It's the worst thing to go through, it changes you as a person. It never leaves you, the worry of flooding again.

Very stressful – it's always hanging over us as a threat now.

Some of these women also made a point of highlighting their attempts to adopt a positive attitude:

I coped by being positive and physically just got on with the repairs.

No one was killed in the floods, and it was just 'stuff' ie. sofas, furniture, carpet, etc. So our outlook was 'it could have been much worse, let's just get on with it'.

Stress, arguments with each other, wanting to cry when out shopping, when really we were very lucky. No one had lost a life, worse things happen to people who have nothing. We at least had a home to go back to.

In Akerkar and Fordham's (2017) research on flooding in the UK, they found mental health effects to be common among men and women. However, women in their study were more likely to feel highly traumatised by the floods, and more likely to say that this was connected to feeling insecure and unsafe in their homes. Their mental burden was also related to the increased caring responsibilities that women were more likely to take on after flooding.

4.7 Flooding, crime and GBV

Overall, 8.7% of respondents reported that they had become a victim of crime during the flooding. This was largely made up of experiences of theft and burglary (7.9%), both in the immediate aftermath of the flooding, and then, for some people, later on when their houses were being repaired:

> *Attempted thefts from property.*
>
> *Vans were coming from other areas and going into homes to take from the upper floors. My husband and neighbours formed a group to monitor our area at night.*
>
> *I witnessed vans of youths seeing if they could break in. Reported to police – no response.*
>
> *Opportunist burglary was a problem with homes being empty for so long.*
>
> *I stayed in my house (one of the very few who did). My husband had to work away so I was very alone. Looters come to our back garden many times.*
>
> *Theft by the builders appointed by the insurers.*
>
> *Shortly after the house restorers started work, approximately £100 was found missing from an upstairs room.*

As well as creating feelings of fear among some of those who stayed in their homes after they were flooded, and some of those in temporary accommodation, having possessions stolen during the flooding can compound the financial consequences of the disaster.

Three women (0.8% of the total sample) reported that they had experienced domestic abuse at the time of the flooding. This is a count of incidence *at the time of the flooding*, rather than prevalence (i.e. those who had experienced domestic abuse before or after the flooding), and it is important to note that domestic abuse is very rarely a one-off crime, but tends to be long term and repeated. In the Methodology section of this chapter (Section 4.4), we suggested reasons why this is likely to underestimate the prevalence of domestic abuse during the floods. Of those three women experiencing domestic abuse, all said that it had happened before the flooding too:

> *Stress on relationships were increased (I had increased difficulties in my relationship with my boyfriend even though he did not live with me).*
>
> *I left my coercively controlling husband in [date] and moved into [street]. I am incredibly lucky to have secured a property under contractual conditions which guarantee*

that I am rehoused should flooding happen again. However I think there is inadequate support for vulnerable women within current flood management strategies. I think this is exacerbated by divorce law in this country, which could better support victims of domestic abuse in leaving their husbands.

In Akerkar and Fordham's (2017) study, also conducted by a survey, a number of respondents reported increased interpersonal conflict and arguments in the post-flooding period, although not violence or abuse.

Our research also provided the opportunity to learn about women's perceptions of the relationship between flooding and domestic abuse, in particular, whether they believe that the two phenomena are connected, and the reasons behind any connection. This is of special interest as this sample of women all have first-hand experience of flooding. The results are mixed, as 34.7% of all respondents said they thought that domestic abuse is likely to increase during flooding, while 31.3% thought this was not likely to be the case (the rest of the respondents did not answer this question, with many of them writing that they did not have enough knowledge to answer).

Sixty-seven respondents commented on their reasons for believing that domestic abuse increases during flooding, with most blaming the multiple stresses involved in experiences of flooding:

I could understand how the high levels of stress and anxiety could lead to this.

Stress makes you act differently and makes people short tempered.

Stress brings out the worst in people. Plus money worries, plus living in reduced circumstances.

Due to homelessness, financial issues, pressure of alternative accommodation, loss of personal belongings and no insurance.

I do believe that there was enormous pressure on families with small children, who were living in very cramped temporary conditions.

It was a very stressful time. For those who suffer from abuse, it would probably get a lot worse. A lot of people in the area turned to alcohol as a result of the flood.

As Enarson et al. (2018) commented in a recent review of GBV during disasters, the idea that stress experienced during disasters 'causes' violence is widespread but misleading. As we go on to argue in Chapter 6, women as well as men are under intense stress when disasters strike, and yet women are much less likely to commit violence. Instead, women's common response is to take on increased caring responsibilities. Instead, the rise in GBV during disasters, whether in low-, middle- or high-income countries, relates to 'strongly felt values and gender ideologies' (Enarson et al. 2018: 208).

Some of our respondents made observations based on their own past experience of domestic abuse, or knew of others who had experienced domestic abuse during the flooding:

> *We coped well as a family but we know others didn't.*
>
> *I informed the police about my elderly neighbour (suspected domestic abuse).*
>
> *I was glad I was on my own. I got on my grown up children's nerves. I did struggle to cope. Within a couple it may have led to abuse.*
>
> *You're restricted as to where you can go [after flooding], so you spend more time together. I have experienced abuse with a previous partner a long time ago. It's not nice.*

On the other hand, the reasons given by those who did *not* think that domestic abuse is likely to rise during times of flooding tended to focus on life being both disrupted and more public, meaning fewer opportunities for abuse:

> *Domestic abuse is usually behind closed doors. During the floods, people moved into hotels, with family or other emergency accommodation. The chance of being heard or seen I would think prevent their disgusting behaviour. My ex husband would always make sure nobody would hear or see before he started!*
>
> *People are removed from normal situations – house and routines – maybe living with others, so less opportunity.*
>
> *I would consider it more likely not [to occur], people would pull together.*

Several women commented that flooding alone would be unlikely to lead to new cases of domestic abuse, if it did not already exist within the relationship:

> *Due to high levels of stress this could escalate in a relationship if this was a problem prior to the floods.*
>
> *If someone is prone to it, then the stress of a flood could make things worse.*
>
> *Anyone who cannot cope with stress, or a controller in a situation he/she cannot control people, prone to taking their anger out on others, would inevitably act on these feelings.*
>
> *I think domestic abuse is part of a pattern of control and it would be only a typical abuse which would increase.*

There is a mixed understanding of domestic abuse, then, from the accounts of the respondents. Those who have past experiences of domestic abuse themselves (a number of respondents referred to experiences that predated the floods) tend to have more accurate perceptions. It is clear from the wider literature on disasters that domestic abuse does increase during flooding; it rarely occurs for the first time in these circumstances, but existing patterns of abuse can become intensified by the difficult social and economic circumstances that families find themselves in. However, whether the family are living in ways that are more public (e.g. with others and/or in temporary accommodation) or does not have a protective effect.

4.8 Conclusion

This study of 380 women whose homes were flooded in England in 2015 shows some similarities and some differences with the two studies from Bangladesh reported in Chapters 2 and 3. Importantly, a much smaller proportion of the respondents in England than in Bangladesh reported that the flooding had significantly worsened their economic circumstances. Although it is important to highlight the fact that severe economic effects were reported by a minority of the respondents in Cumbria, as a whole, the local population was better protected with insurance and government support. As we have discussed in earlier chapters, the increase of poverty after disasters in Bangladesh is one of the factors that is directly responsible for some forms of GBV increasing at those times.

Some of the UK respondents had also been in a more advantageous position as regards being able to move away from the immediate disaster as it unfolded, and then access some form of rehousing for weeks or months afterwards. This is partly due to the lesser scale and reach of the disaster, which affected specific areas within Kendal and Carlisle where the rivers overflowed their banks, but also because women and their families were more likely to be able to stay with relatives or friends or to move to temporary hotels or rental accommodation for which they either paid or received government support. Nonetheless, many women report either staying in or moving back to their damaged, damp and cold homes and then living upstairs for periods of weeks or months while renovations took place, often with consequences for their physical health.

Mental ill health effects were reported to be even more common than physical ill health effects, with large numbers of women in the UK study experiencing stress, anxiety and depression after the floods, a small number having more serious mental illnesses, and symptoms of PTSD reported as widespread and ongoing to the present day. This is broadly comparable to Bangladesh (although women there tend to use fewer clinical terms to describe mental health impacts). The study clearly shows how the trauma of an initial event such as flooding is compounded by the various after-effects, such as the loss of possessions, a worsening of physical and mental ill health, an increase in caring responsibilities, and increases in crime or violence. For some – a minority in the case of the UK study – the disaster becomes layered, with each layer intensifying the experience of the initial disaster. Trauma itself is an important way in which the disaster continues to wind slowly on through time, strongly suggesting the severity of mental health effects when disasters and GBV coincide. Chapter 6 takes up this theme of trauma again.

As regards gender-based violence in the Cumbria flooding, some crime was reported by respondents at the time of the floods, including a very small number of cases of domestic abuse; however, the study was not able to measure the prevalence of abuse as well as its incidence. We also expect some under-reporting due to the methods used for this study. Nonetheless, when detailing the incidents that were of concern to them after the flooding, women were more likely to mention property crime than physical or sexual assault. Some experienced sexual harassment after the flooding, but there is no evidence to suggest this was at higher levels than before the disaster. On the whole, the safety of female victim-survivors of flooding in Cumbria was not reported as being as severely compromised as it was for many of the women in the Bangladesh studies. We go on in Chapter 6 to draw further comparisons with the broader culture in Bangladesh and UK contexts which, we argue, underlie the relationships between flooding and GBV.

The UK study also shows that understandings of domestic abuse during disasters vary considerably, showing some significant gaps in knowledge. Women who have been victim-survivors earlier in their lives generally have a more accurate understanding of whether and why abuse might increase during and after disasters.

Again, there are likely to be aspects of our findings that are local to the settlements in question, and to Cumbria, just as the two studies in Bangladesh reported different gender cultures in the regions of Barguna and Rangpur (Chapters 2 and 3). Akerkar and Fordham (2017: 219) make this case from their study of flooding in two English towns, especially in relation to men and women's post-flood coping strategies which were linked to the 'embedded gender relations and roles in a *place*...[as] traditional forms of masculinity and femininity were predominantly played out in home-neighbourhood places after these floods'. Akerkar and Fordham (2017) concluded, as we do, that gender must be mainstreamed in disaster planning, in high-income as well as low-income countries (see Chapter 7).

References

Akerkar, S., & Fordham, M. (2017). "Gender, place and mental health recovery in disasters: addressing issues of equality and difference". *International Journal of Disaster Risk Reduction* 23: 218–230.

Cumbria County Council (2019). *Domestic Violence Facts*. https://www.cumbria.gov.uk/communitysafety/DomesticViolence/Domestic_Violence_Facts.asp, accessed on 14/01/22.

Cumbria Intelligence Observatory (2015). *Cumbria and Districts 10 Year Population Trends*. Cumbria County Council, Carlisle, UK.

Cumbria Intelligence Observatory (2017). *Kendal ICC Profile*. Cumbria County Council, Carlisle, UK.

Cumbria Police (2019). https://www.cumbria.police.uk/Advice-Centre/Personal-Safety/Domestic-Abuse.aspx, accessed on 14/01/22.

Enarson, E, Fothergill, A., & Peek, L. (2018). "Gender and disaster: foundations and directions for research and practice." In Rodriguez, H., Donner, W., & Trainor, J. E. (eds.) *Handbook of Disaster Research*. Springer, New York, 205–223.

Environment Agency (2016a). *Carlisle Flood Investigation Report: Flood Event 5th-6th December 2015*. Cumbria County Council, Carlisle, UK.

Environment Agency (2016b). *Kendal Flood Investigation Report: Flood Event 5th-6th December 2015*. Cumbria County Council, Carlisle, UK.

Her Majesty's Inspectorate of Constabulary (HMIC) (2015). *Increasingly everyone's business: a progress report on the police response to domestic abuse*. https://www.justiceinspectorates.gov.uk/hmicfrs/wp-content/uploads/increasingly-everyones-business-domestic-abuse-progress-report.pdf, accessed on 07/04/22.

Johnston, I. (2021). "UK risks 'turning clock back' on gender equality in pandemic." *The Guardian*, 9 February 2021.

Little, J. (2017). "Understanding domestic violence in rural spaces: a research agenda." *Progress in Human Geography* 41(4): 472–488. doi:10.1177/0309132516645960

Masson, V. L., Lim, S., Budimir, M., & Podboj, J. S. (2016). *Disasters and Violence against Women and Girls. Can Disasters Shake Social Norms and Power Relations?* Overseas Development Institute, London.

Office for National Statistics (2020a). *Domestic abuse in England and Wales overview: November 2020.* ONS, London.

Office for National Statistics (2020b). *Sexual offences in England and Wales overview: year ending March 2020.* ONS, London.

Otto, F. E. L., van der Wiel, K., van Oldenborgh, G. V., Philip, S., Kew, S. F., Uhe, P., & Cullen, H. (2018). "Climate change increases the probability of heavy rains in Northern England/Southern Scotland like those of storm Desmond – A real-time event attribution revisited." *Environmental Research Letters* 13(2).

Parkinson, D., Lancaster, C., & Stewart, A. (2011). "A numbers game: lack of gendered data impedes prevention of disaster-related family violence." *Health Promotion Journal of Australia* 22(4): 42–45.

Phillips, S. (2011). *Domestic Violence on Rise in Christchurch after Earthquake Stress suite101. com: Insightful writers. Informed readers*, 1 March *2011).* http://www.whealth.com.au/documents/environmentaljustice/women-disaster-violence-litreview.pdf

Schumacher, J. A., Coffey, S. F., Norris, F. H., Tracy, M., Clements, K., & Galea, S. (2010). "Intimate partner violence and Hurricane Katrina: predictors and associated mental health outcomes." *Violence and victims* 25(5): 588–603.

Whitehaven News (2019). "County's domestic abuse has doubled." 18 December 2019, County's domestic abuse has doubled, Whitehaven News.

Womens Policy Group Northern Ireland (2020). *Covid-19 Feminist Recovery Plan.* https://wrda.net/wp-content/uploads/2020/07/WPG-NI-Feminist-Recovery-Plan-2020.pdf, accessed on 07/04/22.

World Health Organization (2021). *Factsheet: Violence against Women.* https://www.who.int/news-room/fact-sheets/detail/violence-against-women, accessed on 14/01/22.

5

DISASTER PATRIARCHY IN HYPER-ISOLATION

Covid-19 and gender-based violence

Rachel Pain

5.1 Introduction

This chapter turns to look at our fourth and final case study, the major global disaster of recent times – the Covid-19 pandemic. The pandemic began shortly after we had completed the other studies that we have presented in Chapters 2, 3 and 4 of this book. Within weeks of its onset, and more than any other disaster, the Covid-19 pandemic had made the connection between disasters and gender-based violence (GBV) highly visible, and pushed the connection between disasters and GBV more firmly onto public and policy agendas, perhaps more effectively than any other event. Significantly, this has taken place globally. There have also been some unique dimensions to the devastating experiences of GBV that have accompanied the pandemic, as perpetrators have actively used the situation to control victim-survivors. Moreover, sources of support, whether formal or informal, have been much harder to access. The pandemic has seen some new and renewed responses to GBV around the world, as this new layered disaster has unfolded, although that is not to say that these have been adequate, or are necessarily likely to result in sustained changes (see Chapter 7).

The chapter examines these issues, first reviewing evidence from around the world on the impact of Covid-19 on GBV. As it shows, the increase in domestic, sexual and other forms of violence against women and girls is just one of the gendered impacts of the pandemic that have fallen disproportionately on women. Secondly, it assesses the social stratification of these added risks of violence, asking which women and girls are most at risk during the pandemic, and what this tells us about the intersectional nature of risk. Thirdly, it analyses the rise in GBV during Covid-19 within a conceptual framework that has two parts to it. Focusing on the distinct spatial aspects of the pandemic, it explores the amplification of intimacy that produces hyper-isolation for victim-survivors. Drawing on Ensler (2021), it then

DOI: 10.4324/9781003089780-5

considers the fit of Luft's (2016) concept of racialised disaster patriarchy (introduced in Chapter 1) to the rise of GBV during the pandemic.

5.2 Covid-19 and gender-based violence: global impacts

The Covid-19 disaster has had distinct spatial impacts on populations, and these have undoubtedly aggravated the rise in GBV. In Chapter 1, we explained what we mean by the 'layering' of disasters; the accumulation of the initial disaster and the other forms of harm that often follow, such as GBV. With many disasters, a significant factor in this layering is that disasters lead to the mass displacement of people from their homes, both immediately and in the longer-term fallout from the disaster, and particularly where disasters are repeated. We witnessed this in the case of many of the victim-survivors of cyclones and flooding in Chapters 2, 3 and 4, both in Bangladesh and the UK. The Covid-19 pandemic has also led to displacement in some parts of the world and exacerbated the other causal factors in displacement, and the virus has also severely affected people who were already displaced by other humanitarian emergencies (Khouzam and Verma 2020). However, for many millions of other people, Covid-19 has been a disaster that does the opposite: restricting people to their homes and increasing the amount of time they spend there with families and others they co-habit with (World Health Organization 2020). The result is that domestic abuse in particular, but also other forms of GBV including sexual assault and child abuse, have become more severe and frequent during Covid-19 outbreaks (Ertan et al. 2020). Meanwhile, the severe economic impacts of the pandemic, particularly in lower-income countries, have led to increased child marriage and trafficking (Appiah 2021); and according to Ensler (2021), dowry violence, female genital mutilation, digital sexual abuse and women going missing have also increased during the pandemic.

Domestic abuse was being described as 'the shadow pandemic' by Phumzile Mlambo-Ngcuka, Executive Director of UN Women just a few months into the pandemic in the spring of 2020. The UN's research showed that rates of GBV had risen for every country for which data were available. The following month, Patricia Scotland, Secretary-General of the Commonwealth, warned of a 'hidden pandemic' taking place, evidenced by

> ...surging numbers of emergency calls to [domestic abuse] helplines – with rises of anything between 25 and 300 per cent, dramatic increases in internet searches for support for those affected by domestic violence, and higher numbers of domestic homicides.
>
> *(Scotland 2020)*

Despite these dramatic early warning signs, data on GBV – which is always difficult to estimate – became even trickier during the pandemic (Sutrich 2020). Data on rates of GBV during the pandemic reflect only those cases which come to attention, and it is not yet known what impact the pandemic has had on reporting

(Walklate et al. 2020). Many victim-survivors never report violence to the authorities at any time and may have been deterred further still because of factors such as lockdowns, fear of catching the virus, and the belief that achieving justice is unlikely during the pandemic.

However, there are now many studies which document rises in GBV, not only in the very early days of the pandemic but in the months that followed. For example, in Bangladesh, there were dramatic rises in domestic and dowry-related violence (Ehsan and Jahan 2020). In the UK, according to the domestic abuse charity Refuge, calls to the Domestic Abuse National Helpline increased by 49%, and in the first two weeks of lockdown domestic homicides doubled, including the deaths of children (BBC News 2020). The Office for National Statistics (2020) reports that the number of arrests for domestic abuse-related crimes between 1 April and 30 June 2020 increased by 24% compared with the same period in the previous year.

In Jingzhou in China, the number of domestic violence reports to police in the early weeks of the pandemic was three times the level of the same period in the previous year (World Health Organization 2020). Similarly, in the US, stay-at-home orders coincided with a rapid increase in GBV (Kofman and Garfin 2020). In many states, calls to helplines and support services increased by between 20% and 62% in the first month. However, some telephone helplines reported reduced call volumes, possibly because of the difficulty of help-seeking in this way when in close proximity to the perpetrator. Moreover, as the use of health services decreased generally during the pandemic, fewer victim-survivors of domestic abuse sought medical care for their injuries, which had previously been a key method of help-seeking to leave abusive partners. Meanwhile, refuges and shelters, which were already at capacity pre-pandemic, came under even more pressure (Kofman and Garfin 2020).

Sutrich (2020, np) highlights a similar paradox from research on sexual violence in several countries during the pandemic. First, the global increase in sexual violence during the pandemic presents 'a clear and increased risk of violence, neglect, sexual exploitation and abuse, discrimination and social exclusion as a result of the compounding impacts of the virus and its secondary effects'. At the same time, services have been disrupted, women may be less able to travel to them due to local restrictions on movement and resources have been diverted to fight the pandemic itself, so that 'the space for survivors to seek care and disclose their experience is shrinking'. Like many other commentators, Javid and Chattu (2020) observe that GBV was already a global pandemic that affected 1 in 3 women and girls long before Covid-19. Under lockdown, they estimate that 15 million GBV incidents were taking place every three months around the world. GBV is also associated with higher mortality and morbidity for co-occurring diseases during the pandemic. These associations are not a new phenomenon, but a known outcome of previous pandemics; for example, in various African countries, Ebola outbreaks were also accompanied by increases in domestic abuse, rape, and teenage pregnancies. And yet GBV has not been prioritised during the latest pandemic in many countries of the world.

The surge in GBV is expected to last several years, as has been documented with previous rises in GBV linked to other disasters. This is because of the medium- and long-term impacts of the pandemic causing ongoing disruption. Increases in trauma and other mental and physical health effects of GBV linked to the Covid-19 disaster are also expected (Kofman and Garfin 2020). As Watson (2020) comments, for the most sharply affected groups, Covid-19 and its associated harms will become a collective trauma which is passed down over time and affects how groups see themselves and each other. This exacerbates the traumatic effects of smaller-scale disasters (see Chapter 4).

These increases in violence, especially against women and girls, comprise just one of the gendered impacts of the pandemic that have fallen disproportionately on women. In many countries, women have been more affected than men, economically, socially and in terms of their health. For example, school closures increased women's duties in particular, as the burden of homeschooling and childcare often fell upon them (World Health Organization 2020). Swaminathan (2021) charts further effects on women and girls' social wellbeing, isolation, income and educational attainment in India. Globally, an estimated 11 million girls may never return to school after the pandemic, owing to the added financial pressures families are under, and this combination leads to higher rates of child marriage, early pregnancy, domestic abuse and trafficking (Appiah 2021). The Covid-19 pandemic has also highlighted the value and importance of care work, whether paid or unpaid, and retail work, which are both predominantly low-paid jobs done by women that place them at significantly higher risk of catching the virus (Dowler and Bartos 2021; Womens Policy Group 2020). Dowler and Bartos (2021) argue that the pandemic has exposed a crisis of care, leading to the failure of medical systems in parts of the US, making those who give and need care most vulnerable to Covid-19. Thus, as we saw in earlier chapters, GBV is one gendered impact of the disaster, entwined with many of its other harmful effects on women and girls.

5.3 The social and regional stratification of the risks of GBV during the pandemic

Despite these broad patterns, that have meant Covid-19 generally impacts more on women than men, there are, of course, highly significant differences between regions and social groups that also shape the relationship between the pandemic and GBV. In this sense, too, it is not only a layered disaster but a slow one (see Watson et al. 2020), with harm caused and reiterated over time by longstanding deeper structures and inequalities rather than the virus itself. Covid-19 has deepened existing inequalities in the same way that GBV does (UN Women 2020) and as we have seen in earlier chapters of this book.

As Eaves and Al-Hindi (2020: 133) put it, 'the Covid-19 pandemic has laid bare material vulnerabilities across space and scale': those people who are the most disadvantaged in economies that are going into recession, those with the poorest health care access, and those with identities and in living conditions that intensify vulnerability have been most affected by the disease. All this is sharply reflected geographically. For example, the particular localities in the US with the highest rates of Covid tend to be low-income and/or communities of colour – communities which are sometimes pathologised as behaving in a way that invites the virus, but in reality which experience underlying structural inequalities and injustices that leave them more vulnerable. In the UK, Bambra et al. (2020) note that similar patterns draw attention to 'the syndemic pandemic', where those areas already experiencing social and economic inequalities have been significantly more affected.

These inequities are also reflected on a global scale. Appiah (2021) describes the global disparities in Covid's impacts as

> A tale of two pandemics...the main engine of destruction wasn't this new disease, but its second-order effects.
>
> *(Appiah 2021, np)*

As he explains, richer and poorer nations differ in their vulnerabilities. While the chief costs of the pandemic in higher-income countries were the lives of older people, in lower-income countries, whose populations have a much younger profile, the main costs were economic, leading to increased morbidity and mortality from a range of other diseases. HIV, malaria and TB were less well managed because of the impacts of Covid-19 on healthcare, and so death rates have risen among the young. Moreover, most importantly, lower-income countries are less able to borrow money to protect their workforce in the way that higher-income countries have done, resulting in a high toll of job losses and people being unable to sell their goods, increasing poverty and impacting the poorest people the most. This has been exacerbated by disruption to patterns of global trade and lower-income countries' dependence on higher-income countries as markets. And as we saw in the case of Bangladesh in Chapters 2 and 3, when poverty increases because of disasters, so do a range of forms of GBV.

As we have already noted in this chapter, some women are far more affected by the pandemic than others, especially low-income women, Black and racialised women, women with disabilities, women with caring responsibilities, LGBTQ women, migrants, domestic workers and sex workers (Ensler 2021; Kofman and Garfin 2020; Sultana 2021; WPG 2020). Women's jobs have been very vulnerable to the effects of the pandemic; in the US, 5 million women lost their jobs, and in India, it was almost 40% of women (Ensler 2021). Sultana (2021) describes how the pandemic has increased discrimination and GBV against the sub-groups of women mentioned above, at the same time bolstering privileges based on class and whiteness. For those already more at risk from GBV, the pandemic was disproportionately devastating. Mwendwa (2020) takes the case of women with disabilities, who are already four times as likely as other women to suffer sexual violence, and among those most affected by pandemics. Indeed, the Covid-19 pandemic exacerbated cases of sexual violence against women with disabilities, as people were quarantined with carers, less able to access support services, and, in some cases, wanted to avoid shelters which can be unsafe for those most vulnerable to the virus.

All of this strongly suggests that an intersectional approach to GBV is needed during Covid-19 and future pandemics, which recognises the additional risks and challenges experienced by women and girls due to race, ethnicity, sexual orientation, disability, age and other inequalities (Pentaraki and Speake 2020).

5.4 Hyper-isolation during the pandemic

GBV may have become more visible during the Covid-19 pandemic, but the causal mechanisms underlying the relationship are not necessarily well understood. (In Chapter 4, our analysis of respondents' beliefs about the connection between domestic abuse and flooding came to a similar conclusion.) In particular, Covid-19 has often been scapegoated as the cause of GBV in popular, academic and policy discourses (Pentaraki and Speake 2020). In these representations, rises in violence during the pandemic have been blamed on increased stress, frustration and aggression, rather than the abuse simply being a continuation and escalation of violence that was present in 'normal' times. Home is always the most dangerous place for women and children, pandemic or no pandemic (Walklate et al. 2020), and

> It is important to remember that Covid-19 and its related stressful conditions do not transform people who have no inclination to control their partner into perpetrators.
>
> *(Pentaraki and Speake 2020: 204)*

Nonetheless, there are particular features that contribute to the rise in GBV during this pandemic that make it quite unique among disasters. During the pandemic, families spent much more time together than usual in the private space of their homes. Intimacy was therefore intensified. It was not the virus or disease itself that aggravated GBV, but rather the government responses that required lockdown, self-isolating, quarantine and social distancing had a number of effects.

Much of the rise in domestic abuse after the pandemic began was exacerbated by government instructions to lockdown at home for many weeks, which created 'a perfect storm for controlling, violent behaviour behind closed doors' (UN Women 2020). Not only were many victim-survivors far more exposed to perpetrators in terms of time and proximity, but they had correspondingly less contact with family and friends who previously may have provided informal support (World Health Organization 2020). Help-seeking is difficult at the best of times but almost impossible when perpetrators are ever-present (Drew 2020); in non-pandemic times, those experiencing domestic abuse usually reach out for support when they are outside the home, or by phone or internet when the perpetrator leaves the house. In addition, leaving an abusive partner for a refuge or other 'safe' place risks greater exposure to the virus, which Kofman and Garfin (2020) describe as presenting 'a unique and distressing paradox for victims'.

Moreover, there is much evidence from around the world that perpetrators have deliberately used Covid-19 restrictions to exercise control over their partners, and to isolate them further from sources of support (Pentaraki and Speake 2020; World Health Organization 2020). In the US, Kofman and Garfin (2020) report domestic abusers using coercive control to harness Covid-19 and instil fear and compliance in their partners. The conditions created by lockdown measures and social distancing therefore overlap with perpetrators' existing strategies for control (Van Gelder et al. 2020). Some have effectively used the virus as a weapon, by blocking victim-survivors' access to medical treatment, or giving misinformation about lockdown or quarantine measures (Pentaraki and Speake 2020). Women are also more at risk from economic abuse during times of economic hardship such as that created by the pandemic (World Health Organization 2020), and in many parts of the world, job losses, a rise in poverty and bereavement have intensified this abuse (Appiah 2021).

Together, then, the typical controlling tactics of GBV perpetrators combine with the Covid-19 pandemic to create conditions that we might call *hyper-isolation*. Isolating itself – the practice of gradually cutting off victim-survivors from their support networks, and, in extreme cases, the outside world in its entirety – is a longstanding tactic of abusers that is present across cultures and time (Pain 2021; Sahdan et al. 2021; Stark 2009). Elsewhere, I have suggested that

> The isolation of GBV survivors is foregrounded in decades of feminist scholarship on GBV...Isolation is not just 'how it is', but a condition created and exploited by perpetrators, and buttressed by social perceptions and practices, so that isolation is normalized and isolating unseen.
>
> *(Pain 2021, np)*

Isolating is commonly accompanied by persistent attacks on the victim-survivor's personality, capacities and actions, with the long-term effects of eroding self-confidence, negatively affecting mental health and causing trauma. In the same paper from which the quote above comes, I analysed survivors' experiences of four forms of isolating that are purposeful effects of physical, sexual and psychological GBV. First are the physical restrictions imposed by perpetrators, which lead to the survivor's world contracting. Secondly, psychological isolating is a well-established tactic of perpetrators, involving the breaking down of their targets' psyche so that they are less able to gather resources to challenge violence. Third are the processes of material and social isolating by which victim-survivors who try to leave their perpetrators are often punished with the loss of homes, jobs, education and contact with friends and family, and sometimes their children. And fourth are isolating societal discourses about GBV, which tends to be positioned in popular beliefs as a problem of individuals or dysfunctional families, and related to the characteristics of victims rather than perpetrators. The isolation that perpetrators foster is sometimes reinforced by what happens when victim-survivors try to seek help, especially those who are already marginalised or stigmatised in the cultures in which they live. While isolating is a tactic of individual perpetrators, then, it is also a structural condition, and the formal and informal responses by others to abuse may have the effect of propping it up rather than challenging it (Pain 2021).

The Covid-19 pandemic adds another layer to these isolating effects, making violence and the strategies of coercive control that sustain it more severe and more frequent. At the same time, the public health strategies of isolating and distancing during the pandemic mean that GBV becomes harder to resist and survive (Ertan et al. 2020). So while rising rates of GBV during the pandemic tend to be represented as a result of lockdown itself, women's organisations have asserted that these are unlikely to be completely new cases but result from more isolation, more exposure to perpetrators and therefore more danger. Covid-19, or rather responses to it, have led to an intensification of abuse, as the 'intimate captivity' that perpetrators foster (Herman 2015; Stark 2009) is reinforced.

In the UK, at the end of March 2020, the Home Secretary announced that those experiencing domestic abuse were exempt from lockdown restrictions if they left home to seek help from refuges or the police. However, this did not help them to find a way to leave when being victimised by partners who were constantly at home and watching them, or improve their chances of being granted a place by a struggling refuge sector that had contracted after a decade of austerity measures. In this way, the isolating effects of GBV have also been compounded by the enclosure of support services under conservative fiscal policies (Pain 2021), heightening risks further during the pandemic. Emergency funding was made available to support services by May of the same year, to address the increase in demand and some of the complexities of service provision under lockdown, but the sector remains perilously underfunded.

5.5 Racialised disaster patriarchy and the pandemic

In Chapter 1, we introduced the concept of racialised disaster patriarchy, drawing on Luft's (2016) argument that not only do gendered patterns help to produce disasters, but that '*part of what is disastrous* is patriarchy' (Luft 2016: 8, our emphasis). In short, responses to disasters reinforce patriarchy and the conditions that make women and girls particularly vulnerable to disasters in the first place, and this process is always racialised, reflecting the intersections of racism and patriarchy. Luft (2016) evidenced this in her in-depth analysis of state and NGO responses to Hurricane Katrina in the US, demonstrating how racism and sexism worked together to produce *both* the disaster *and* the responses of formal disaster management activities and progressive social movements, exacerbating the disaster for women and people of colour. This concept of racialised disaster patriarchy moves beyond the identification of gendered experiences, to examine how gendered assumptions, discrimination and oppression are part of the whole lifecourse and underlying logics of disasters.

Ensler (2021, np) has recently used Luft's framework to argue that the Covid-19 crisis has also been exploited in ways that reassert patriarchal control and dominance, 'on the one hand, escalating the danger and violence to women, and on the other, stepping in as their supposed controller and protector'. From interviews with activists and grassroots leaders around the world during the pandemic, Ensler identifies five key factors that compound the crisis for women. Women become more unsafe, as is evidenced by the rise in GBV. They lose economic power, evidenced by the job losses mentioned earlier in this chapter. They lose autonomy, and education, and finally 'they are pushed onto the frontlines, unprotected, to be sacrificed', due to their predominance as workers in the health, social care and education sectors. Both mental and physical health effects have been documented associated with these heavy caregiving roles during the pandemic, and frontline workers have often been women who were inadequately protected with equipment. In line with Luft's (2016) analysis of Hurricane Katrina, Ensler finds that racialised women have been most affected by all of this during the pandemic. She concludes,

> Covid has revealed the fact that we live with two incompatible ideas when it comes to women. The first is that women are essential to every aspect of our life and our survival as a species. The second is that women can easily be violated, sacrificed and erased. This is the duality that patriarchy has slashed into the fabric of existence, and that Covid has laid bare. If we are to continue as a species, this contradiction needs to be healed and made whole.
>
> *(Ensler 2021, np)*

Others concur that it is the shape of Covid-19 policy-making, rather than Covid-19 itself, that has fuelled negative impacts on women including the rise in GBV. Swaminathan (2021) reports from South India, where:

> The implementation of crucial, but often blunt, public health measures such as stay-at-home policies and the disruption of key services like schools and health facilities...have increased violence against many women and girls.
>
> *(Swaminathan 2021, np)*

Operating in crisis mode meant there was no pause to predict or monitor these highly gendered effects. Javid and Chattu (2020), quoting UN chief Antonio Guterres, state that the pandemic has exposed what we already knew – that there is a male-dominated culture around the world that is damaging to everyone. Pandemic economic and social policies have failed, leaving behind fragile healthcare, social protection and justice systems that failed to protect many during the pandemic. For Watson et al. (2020), Covid-19 has been a slow disaster, as the conditions that have exacerbated it have gathered for decades without notice: poor housing, race inequality, health inequalities and poverty. Jacobs (2021) names Covid-19 as a disaster of racial capitalism; initially a crisis, it only became a disaster because state responses turned it into one. She argues that racial capitalism makes Black, Indigenous and Latinx people highly vulnerable in the US. These groups experienced significantly higher mortality rates during the pandemic, also reflected in other parts of the world because of intersecting structures of race, class, gender and capitalism.

5.6 Conclusion

The Covid-19 pandemic provides a compelling case study of the direct connections between disasters and GBV. The pandemic is expected to continue to have impacts for years to come, and it is likely that we will not fully understand its impacts on different forms of GBV for some time (Drew 2020). This chapter has presented a range of evidence about GBV and questioned why it has appeared to rise so dramatically in different parts of the world after the onset of the pandemic. In all parts of the world, these increases were evident, accompanied by significant reductions in victim-survivors' access to help and support, and the longer-lasting economic and social impacts that are also likely to increase GBV in the future.

The pandemic very effectively illustrates our broader argument in this book that disasters are layered. Covid-19, government and community responses to it, different forms of GBV, and the further impacts of economic crises and rising poverty all compound each other. Addressing the root causes of the pandemic, Sultana (2021) argues that climate change and Covid-19 are interrelated crises that reflect and reinforce structural inequalities and injustice. We have made a related argument in relation to the connections between cyclones, climate change, poverty and GBV (Rezwana and Pain 2020).

Rather than accepting at face value the idea that the stress of lockdowns, alone, has aggravated GBV during the disaster, the analysis here has identified particular processes operating in intimate spaces, wherein regulations introduced as pandemic responses have effectively supported longstanding tactics of perpetrators. Evidence suggests that perpetrators of GBV have actively used isolation as part of their coercive control, to commit domestic abuse within the home. Outside the home, other forms of GBV such as child marriage, dowry violence and trafficking have been amplified by the huge economic pressures that have accompanied the pandemic – not only arising from lockdowns and temporary employment closures but from the disruptions to global trade that disproportionately affect lower-income countries.

Furthermore, the chapter has shown how Luft's (2016) concept of racialised disaster patriarchy applies to pandemic responses in which the disproportionate impacts of Covid-19 on women are under-acknowledged (Ensler 2021). Some formal responses have actively increased the vulnerability of women and girls, as well as ethnically and racially minoritised communities, migrants, people with disabilities and particular workers such as carers. As we noted in Chapter 1, these so-called 'vulnerable groups', including women, make up a majority of the population in most countries, and they should not be considered special cases whose needs can be addressed *after* the main business of disaster management and planning has been attended to (Seager 2014). They should be the main focus of effective responses and fully participate in developing them, in the Covid-19 pandemic and in disasters yet to come. Chapter 7 will pick up on this theme. First, Chapter 6 draws together the evidence about disasters and GBV from across the case studies we have looked at so far.

References

Appiah, K. A. (2021). "A tale of two pandemics: the true cost of Covid in the global south." *The Guardian*, 23 November 2021.

Bambra, C., Riordan, R., Ford, J., & Matthews, F. (2020). "The COVID-19 pandemic and health inequalities." *Journal of Epidemiological Community Health* 74: 964–968.

BBC News (2020). "UK lockdown: calls to domestic abuse helpline jump by half." www.bbc.co.uk/news/uk-52433520, accessed on 04/05/2022.

Dowler, L. & Bartos, A. (2021) *Displacing Care* (paper presentation). AAG Annual Meeting, 2021, Seattle, U.S.

Drew, M. (2020). "Feminist perspectives on disaster, pandemics, and IPV." In Geffner, R., White, J. W., Hamberger, L. K., Rosenbaum, A., Vaughan-Eden, V., & Vieth, V. I. (eds.) *Handbook of Interpersonal Violence Across the Lifespan*. Cham, Springer.

Eaves, L & Al-Hindi, K. F. (2020). "Intersectional geographies and COVID-19." *Dialogues in Human Geography* 10(2): 132–136.

Ehsan, S. M. A. & Jahan, F. (2020). "Analyzing the impact of Covid-19 on the mothers of Bangladesh: hearing the unheard." *Research Square*. doi:10.21203/rs.3.rs-79726/v1

Ensler, E. (2021). "Disaster patriarchy: how the pandemic has unleashed a war on women." *The Guardian*, 1 June 2021.

Ertan, D., El-Hage, W., Thierrée, S., Javelot, H., & Hingray, C. (2020). "COVID-19: urgency for distancing from domestic violence." *European Journal of Psychotraumatology* 11: 1. doi:10.1080/20008198.2020.1800245

Herman, J. L. (2015). *Trauma and Recovery*. New York, Basic Books.

Jacobs, F. (2021). "Beyonf social vulnerability: COVID-19 as a disaster of racial capitalism." *Sociologica* 15(1): 55–65.

Javid, S. & Chattu, V. J. (2020). "Patriarchy at the helm of gender-based violence during COVID-19." *Aims Public Health* 8(1): 32–35.

Khouzam, A. & Verma, M. (2020). "Internal displacement and COVID-19: taking stock and looking forward." *Refugee Survey Quarterly* 39(4): 591–601.

Kofman, Y. B. & Garfin, D. R. (2020). "Home is not always a haven: the domestic violence crisis amid the COVID-19 pandemic." *Psychological Trauma: Theory, Research, Practice, and Policy* 12(S1): S199–S201.

Luft, R. E. (2016). "Racialized disaster patriarchy: an intersectional model for understanding disaster ten years after hurricane Katrina." *Feminist Formations* 28(2): 1–26.

Mwendwa, N. M. (2020). "Disability and sexual violence in the Covid-19 era." *Humanitarian Law and Policy Blog*. https://blogs.icrc.org/law-and-policy/2020/12/03/disability-sexual-violence-covid-19, accessed on 07/04/22.

Office for National Statistics (2020). "Domestic abuse during the coronavirus (Covid-19) pandemic, England and Wales: November 2020." https://www.ons.gov.uk/peoplepopulationandcommunity/crimeandjustice/articles/domesticabuseduringthecoronaviruscovid19pandemicenglandandwales/november2020, accessed on 07/04/22.

Pain, R. (2021). "Collective trauma? Isolating and commoning gender-based violence." *Gender Place and Culture*. doi:10.1080/0966369X.2021.1975103

Pentaraki, M. & Speake, J. (2020). "Domestic violence in a COVID-19 context: exploring emerging issues through a systematic analysis of the literature." *Open Journal of Social Sciences* 8: 193–211.

Rezwana, N. & Pain, R. (2020). "Slow violence and layered disasters: gender-based violence before, during and after cyclones." *Disasters* 45(4): 741–761.

Sahdan, Z., Pain, R., & McEwan, C. (2021). "Demonic possession: narratives of domestic abuse and trauma in Malaysia." *Transactions of the Institute of British Geographers*. doi:10.1111/tran.12485

Scotland, P. (2020). "The hidden pandemic alongside Covid-19." https://thecommonwealth.org/news/hidden-pandemic-alongside-covid-19, accessed on 07/04/22.

Seager, J. (2014). "Disasters are gendered: what's new?" In Singh, A. & Zommers, Z. (eds.) *Reducing Disaster: Early Warning Systems for Climate Change*. Dordrecht, Springer: 265–281.

Stark, E. (2009). *Coercive Control: How Men Entrap Women in Personal Life*. Oxford, Oxford University Press.

Sultana, F. (2021). "Climate change, COVID-19, and the co-production of injustices: a feminist reading of overlapping crises." *Social & Cultural Geography* 22(4): 447–460.

Sutrich, S. (2020). "Covid-19, conflict and sexual violence: reversing the burden of proof." *Humanitarian Law and Policy Blog*. https://blogs.icrc.org/law-and-policy/2020/06/19/covid-19-conflict-sexual-violence/, accessed on 07/04/22.

Swaminathan, S. (2021). "Covid-19 is fueling a pandemic of violence against women and girls." *International Science Council Blog*. https://council.science/current/blog/covid-19-is-fueling-a-pandemic-of-violence-against-women-and-girls/, accessed on 07/04/22.

UNWOMEN (2020). "Violence against women and girls: the shadow pandemic." https://www.unwomen.org/en/news/stories/2020/4/statement-ed-phumzile-violence-against-women-during-pandemic, accessed on 05/04/22.

Van Gelder, N., Peterman, A., Potts, A., O'Donnell, M., Thompson, K., Shah, N., & Oertelt-Prigione, S. (2020). "COVID-19: Reducing the risk of infection might increase the risk of intimate partner violence." *EClinicalMedicine* 21: Article ID: 100348.

Walklate, S., Richardson, J., & Godfrey, B. (2020). *Domestic Abuse-Family Violence, Disasters and Restrictions under Covid-19: An Overview*. Working Paper No. 1. Domestic Abuse: Responding to the Shadow Pandemic. University of Liverpool.

Watson, M. F., Bacigalupe, G., Daneshpour, M., Han, W.-J., & Parra-Cardona, R. (2020). "COVID-19 interconnectedness: health inequity, the climate crisis, and collective trauma." *Family Process* 59: 832–846.

Womens Policy Group Northern Ireland (2020) *Covid-19 Feminist Recovery Plan*. https://wrda.net/wp-content/uploads/2020/07/WPG-NI-Feminist-Recovery-Plan-2020.pdf, accessed on 07/04/22.

World Health Organization (2020). *COVID-19 and violence against women: what the health sector/system can do*. WHO, Geneva.

6

ANALYSING THE RELATIONSHIPS BETWEEN GENDER-BASED VIOLENCE, DISASTERS AND PLACE

Nahid Rezwana and Rachel Pain

6.1 Introduction

Chapters 2–5 have presented case studies of GBV during cyclones and flooding in different regions of Bangladesh; trauma and GBV during flooding in the UK; and GBV during the Covid-19 pandemic across the world from 2020. We now ask what we can learn from analysing these different accounts of layered disasters, as we referred to them in Chapter 1, across different geographical and cultural contexts. What are the commonalities, what are the differences, and what implications does this have for theorising and responding to GBV, before, during and after disasters? This penultimate chapter therefore draws together the findings from the book so far, in an attempt to forge a deeper analysis of the relationships between gender-based violence (GBV), disasters, place and culture.

First, the chapter discusses the finding that GBV increases alongside disasters across the world (as we saw in Chapter 1), but that its forms, severity and impacts are closely related to aspects of location and place. At the micro-level, it differs within countries (e.g. within Bangladesh between the examples in previous chapters of Rangpur and Barguna) and at the macro-level between countries (Bangladesh and UK). Moreover, the global Covid-19 pandemic has provided a horrifying case study of these differences, as GBV has risen across the world. While we discuss the local forces shaping patterns of GBV, we also highlight the importance of global structural processes.

Second, the chapter goes on to examine the trauma that arises from disasters and GBV, charting some of the long-term psychological impacts of these layered disasters, and showing the material effects of trauma on victim-survivors as they try to rebuild their lives. Their situation may be more or less severe according to where victim-survivors live, their socio-economic status and experience of other inequalities, and the availability of formal and informal support.

DOI: 10.4324/9781003089780-6

Finally, the chapter examines the reasons, factors and contexts for the gendered nature of violence – what the research suggests as to why people become perpetrators and victim-survivors in different times and places, including during disasters. Considering various theories and examples, and tackling some of the myths that exist around the causes of GBV, the chapter describes why men are most likely to become perpetrators, and women victim-survivors, irrespective of geographical location. We argue that GBV is historically and structurally embedded in every society. We also identify cases in which women become abusers or enablers of GBV, and discuss the ways in which gendered analysis can explain these patterns.

6.2 Place and culture in GBV and disasters

Geographical location has a powerful influence not only on the likelihood of certain disasters occurring but also on the nature of disaster impacts. Socio-economic conditions, culture, population density and levels of disaster preparedness and resilience of different populations are all important factors. However, in the case of GBV, which frequently rises during disasters and post-disaster periods leading to layered disasters, as we have argued, gender intersects with these other factors to have a very significant impact on outcomes (Bradshaw and Fordham 2015; Enarson et al. 2018; Luft 2016; Seager 2014). As we described in the previous chapters, victim-survivors' experiences of GBV during the floods in Northern Bangladesh and the UK, cyclones in Southern Bangladesh, and Covid-19 in countries around the world help to understand these complex relationships. The following sections bring this material together by focusing on these experiences of GBV, describing some of the commonalities and differences in the impacts of disasters among diverse communities, and demonstrating the strong influence of gender, culture and place.

Similarities and differences: an intersectional approach

As Chapter 1 detailed, and as the research reported in previous chapters on floods, cyclones and Covid-19 confirm, prevailing conditions of GBV become worse during disaster and post-disaster periods, and women and girls are the most likely to become victim-survivors. The forms and types of GBV in different parts of the world show both similarities and differences. For example, domestic violence and sexual assault are widespread across the world's major regions (World Health Organization 2021), while lower-income countries experience far higher numbers of child marriages and dowry violence.

As our research has shown, child marriage becomes a common coping strategy for families during both flood and cyclone disasters in Bangladesh, used by parents to reduce their economic burden during post-disaster poverty. Child marriage is also seen to lower the dowry money payable on marriage, and sometimes parents hope that it will provide a safer living environment for girls – at least in terms of the physical fabric of their housing. Alternatively, children may be sent away from home to earn money, and trafficking becomes a great threat for girls. Meanwhile, the loss of jobs and livelihoods, and living in damaged housing, increases tensions in households during the post-disaster period, further increasing domestic violence (Abramowitz and Moran 2012; Chanthy and Samchan 2014). In this way, poverty intensifies domestic violence post-disaster. As we have seen, women may be abused to extract further dowry money, or as a punishment for collecting relief. However, in many cases, women are also abused by their husbands for no specific reason, as was reported by GBV victim-survivors in both cyclone- and flood-affected regions in Bangladesh.

We also saw that in areas where disaster shelters provide the only option for most people who are forced to flee their homes due to cyclones or flooding, these shelters are often unsafe places for women and girls. Many reported being subjected to sexual harassment, physical and verbal abuse, and rape or rape threats. They do not feel safe staying in the shelters, which adds to trauma and other mental health problems experienced after disasters. This unsafety also makes them more likely to stay in homes that are damaged or unsafe.

For both the flood and cyclone victim-survivors in Bangladesh, it was more difficult to collect data on GBV from comparatively wealthier women in the areas studied. In Barguna, these women did not want to participate in the survey, and in Rangpur, wealthier victim-survivors could not be reached. The research suggested that GBV is more likely to affect women and girls in poorer socio-economic groups after floods. However, we must treat this finding with caution; as is the case in other contexts, it may be that wealthier women are less likely to make public their experiences of GBV, whether to neighbours, formal authorities, or researchers.

In the Cumbria study, for reasons discussed in Chapter 4, the numbers of reported GBV cases were not high enough to allow meaningful correlation with social class or education of respondents. In the UK, widely held public perceptions of domestic abuse view it as most commonly associated with women from poorer households and areas. However, this is one of many myths associated with domestic abuse, and can largely be explained by lower-income victim-survivors being more likely to have to access support services, because of being less likely to have the financial and social capital that help survivors to independently rehouse themselves and their children. Even in the UK, the stigma of experiencing domestic violence is such that most keep it hidden, and do not bring it to the attention of public services unless they have no other choice. There are also well-justified fears among lower-income women and racially minoritised women that their children may be taken into care if they disclose abuse (Wild 2020). While experiences of GBV may be shared by middle- and working class women, the impacts that GBV has on life chances can be markedly different. As Chapter 5 showed, these differences have been especially stark during the Covid-19 pandemic.

There are also differences between the localities studied in Bangladesh. In particular, the two case studies there that were presented in Chapters 2 and 3 provide an opportunity to examine how local cultures and social practices around gender, gender roles and responsibilities create differences among GBV victim-survivors and the ways in which they are able to cope.

The study on cyclone-affected communities in Barguna reveals that domestic violence against women is very common and severe in this area, and that in some cases, where women had entered higher education or had permanent jobs, this did not make any difference. Overall, after considering the employment status, education, socio-economic condition, religion, health and disability of the respondents, the study revealed that these factors do not protect female GBV victim-survivors from abuse (see Chapter 2). There were women in highly educated groups, women who earnt a regular income from employment, lower middle class women and women of different religions all of whom had been abused. Their experiences of abuse had much in common. What differed, however, is their coping strategies (as we discuss in Chapter 7).

For example, one respondent mentioned the problems of being a woman who had a paid job in Barguna:

> *Earning women face double burdens; the burden of torture and the burden of helplessness for not sharing the problems and pains with others. They mostly keep these problems hidden. They spend every single penny for the family and cannot even see their own earnings. She might sometimes want to spend money at her father's house but it creates conflicts with her husband. He does not allow that. Whereas it is a common concept of the society that women do not listen to their husband when they earn.*
>
> *(Female, Barguna)*

> *I earn, I take care of the family and I get beaten at every step.*
>
> *(Female, Barguna)*

Similar cases were revealed in Mahapatro's (2018) research in which 14,567 women participated from 18 states of India. It showed that women's employment does not help in reducing violence against them; in fact, more working women (49%) were facing domestic violence compared to housewives (36%), especially those who were working as unskilled labourers.

However, the situation for Rangpur women living in the Kashiganj village (see Chapter 3 for more details) is different. Here, higher education and/or having a permanent income do seem to help in enhancing women's empowerment and reducing the rates of domestic violence against them. According to the study, wealthier and educated families rarely experience domestic violence:

> *Educated women and women with good jobs rarely get abused, they get some honour in this village, and especially wealthy families do not have wife beating cases in this village.*
>
> *(Female, key informant, Rangpur)*

As we reported in Chapter 3, girls and women from elite and wealthy families in this village do not experience insecurity because of eve-teasing while walking alone during the daytime. This is quite different for women from the poorer families of the same village.

In most cases, women in Rangpur are permitted to work to earn money, mostly in low-paid jobs, and this seems to be associated with reduced domestic violence against them. They help their families to recover from the effects of floods by working in the brick fields or crop fields, as maids or by migrating to other cities to earn their living. Getting involved in prostitution or extra-marital relationships as ways of reducing flood-induced poverty was also reported during the study in Rangpur. Similar practices are also found among some social groups in India; Dutta (2017) reported that culture and tradition justify some women's involvement in prostitution after natural disasters like floods. Women take responsibility for their families during the post-disaster period, and involvement in this profession can be seen as part of that duty.

On the other hand, the study in Barguna shows that getting permission to work outside the home is not easy for married women. Few are permitted to work, while others work outside the home after they are divorced or abandoned by their husbands, or after losing economic support from their parents.

The local culture and social practices in Rangpur are also different in other important ways, particularly regarding social attitudes to victim-survivors. Generally, here, social norms permit women to express opinions against domestic violence and to provide help and support to other women when they are being abused. Though this does not in itself prevent or stop violence, it is very important in spreading awareness about violence against women, and condemnation of it. Some educated women in Rangpur were found to be actively working to stop young couples from getting involved in domestic and dowry violence. It is possible that these efforts actually produced more shame among victim-survivors, too, as in the fieldwork in Rangpur, it was very challenging at first to reach domestic violence victim-survivors who were willing to talk. People still feel it is a problem which should be hidden, and victim-survivors sometimes wanted to avoid the topic during interviews. However, this was very different in Barguna, where female GBV victim-survivors wanted to talk about their experiences and showed their bruises, and men also wanted to talk about this social problem due to its seriousness. In general, women were less empowered in many ways in Barguna, and yet more open about suffering GBV.

During the fieldwork in Rangpur, however, many people openly discussed child marriage as a social problem. They described how parents arrange marriages for their daughters at a very young age to reduce daily expenditure during flood-induced poverty. In addition, they reported a trend among teenagers to get involved in love affairs and elope before the age of 18, which is not allowed and unacceptable in this society. When this happens, the parents of these young couples arrange their marriage as soon as possible (Horii 2020). As a result, young married couples are easy to find in this village. For some, conflict and domestic violence become a problem in their relationships. Their parents are often active in taking action when this happens, irrespective of their socio-economic background. However, poorer families may also be aware of abuse taking place and want to help their daughters but often cannot take action in time against the developing violence due to lack of financial resources. In Barguna, on the other hand, regardless of the economic status of the young wife's family, social shame and dishonour become strong factors in tolerating violence which is known to be taking place. Either the victim herself, or her family, generally do not want to take action in getting a separation or divorce, or filing complaints against the abusive husband and his family, unless the violence becomes very severe.

Place, socio-economic status and other social inequalities work together to produce differential outcomes of GBV before, during and after disasters. This underlines the importance of the intersectional framework that we proposed in Chapter 1 (see Box 1.3), which also underpins the next section.

Place matters

As the above discussion demonstrates, GBV may be endemic across the world, but aspects of local culture and place can strongly influence GBV and its outcomes, as well as the opportunities for victim-survivors and those around them to challenge it. Culture and social practices around gender, gender relations and responsibilities thus create differences between different societies as regards the forms, types and severity of GBV (O'Brien and Macy 2016). The two Bangladesh studies reported in this book are located in the north and south of Bangladesh, with a distance of only 550 km between them, but experience clearly visible differences in GBV conditions and social attitudes about these incidents. Here, place-based analysis helps to understand how particular locales are clearly different from one another considering their cultural or economic conditions, remoteness from the capital, and urban or rural characteristics of the area (Pruitt 2008; see Box 1.3). In the UK, Akerkar and Fordham (2017) studied two floods in Tewkesbury, Gloucestershire, in 2007 and Morpeth, Northumberland, in 2008, and revealed that embedded gendered relations and roles in the two places, including traditional forms of masculinity and femininity in home-neighbourhood places influenced the coping and recovery strategies of men and women. Another piece of research showed how even in a small country like the UK, there are large regional differences in the number of domestic homicides (Smith et al. 2012 in Payton et al. 2017).

Again, analysing the occurrence of intimate partner violence (IPV) at a larger scale shows that it is higher in some parts of the world like the East Mediterranean region, South-East Asia and Africa which has rates of more than 36% (Krahé 2018). Decker et al. (2015) conducted an age-place interaction in IPV risk among low- and middle income countries, finding that broader social and political environments influence gender- and age-related vulnerabilities. Social norms, women's social status, laws and practices influence women's vulnerability to IPV, whereas age-related vulnerabilities are reinforced by relative inexperience and limited social standing. For example, Bangladesh is distinct for having a high level of child marriage (UNICEF 2021). However, social and political unrest in other places may overshadow age-related vulnerabilities to IPV, for example, war-affected regions in Africa (Decker et al. 2015). The impact of culture and place is also clear in Dutta's paper, where she revealed that trafficked women or women living in brothels in India belong to particular social groups from disaster-prone regions of India (Dutta 2017).

Poverty, structural violence and place

Domestic violence against women is very common in the most disaster-prone societies, and these are also often highly vulnerable to poverty. Fluri and Piedalue (2017) argue in their paper that narrow presentations of race and culture do not explain 'the intertwined geographies of corporeality and violence' (p 534), rather that

> Providing a geographic perspective on gendered violence, race, and culture makes visible linkages between intimate forms of violence and systemic/institutional/structural/state violence.
>
> *(Fluri and Piedalue 2017: 539)*

Pruitt (2008), too, argues that domestic violence is not 'truly a local issue', but that the global economy has a strong influence on domestic violence in rural areas by creating conditions that increase vulnerability and risk. Montesanti (2015) also prioritises these forms of structural violence, which are marked by unequal access to the basic needs (housing, healthcare, employment, education), enhancing conditions for interpersonal violence and gendered forms of violence against women. Remote geographical locations and disaster-affected agricultural economies enhance the structural inequalities of capitalism in the regions studied in this book (Barguna and Rangpur in Bangladesh, and Cumbria in the UK). These factors work together to explain persistent poverty and lack of livelihood and work opportunities in those areas, which along with gendered cultures and social practices exacerbate domestic violence. For example, different forms of GBV are more likely to occur in shelters and damaged houses compared to stronger houses in a disaster-hit region. Women running for safe shelter during the warning period, or staying at the shelter, may become victims of sexual harassment irrespective of education and income status. Place, economic conditions and the inclusion of these considerations in disaster preparedness have important consequences for experiences of GBV on the ground.

We argue, then, that rather than having an individualised view of 'place' which involves only local sets of influences on GBV, a framing for understanding the impacts of GBV during disasters must account for systemic inequalities and the sanctioning of abuse within wider sets of social and political relations (Fluri and Piedalue 2017; see also Chapter 1). These systemic inequalities and wider sets of social and political relations affect places differently. As we argued in Chapter 1, layers of structural violences, some of which also exacerbate and result from disasters, coincide and come together in particular places. These variations can, of course, be seen as a source of hope, for example, as women support each other and find ways to fight to change structural injustices (see Chapter 7).

This variable patterning of disaster patriarchy (Luft 2016; see Chapter 1) is underlain by the fact that, irrespective of geographical location, socio-economic conditions and local culture, GBV increases during and after disasters. Some argue that 'violence doesn't have a race, a class, a religion, or a nationality, but it does have a gender' (Solnit 2014: 21) and that violence itself is 'a distinctly gendered component' (Kilmartin 2015: 15). Yet race, social class and histories of colonisation intersect in important ways with gender. According to Dustin (2016), when violence against women occurs among minority groups, their culture is often blamed, and violence may be named depending on the victim-survivor's ethnic origin and culture. Whereas the same kind of violence against women among white populations may be named differently, and 'it is not identified as a problem of either UK, white or western culture' (Dustin 2016: 56). Nonetheless, masculinity, 'a reaction against passivity and powerlessness, and with it comes a repression of a vast range of human desires and possibilities: those that are associated with femininity' (Kaufman 1987: 486), is the root cause of violence against women, though rarely identified in the media, political discourse or court rulings. Section 6.4 will pick up on discourses of masculinity, and the ways in which perpetrators or victim-survivors are created within society and the family, shaped too by culture, gender and place.

6.3 The role of trauma within layered disasters

In this section, we return to the important issue of trauma. We introduced the role of trauma in relation to disasters and GBV in Chapter 1, and then in the context of the Cumbria case study in Chapter 4. Our suggestion is that trauma is not just an after-effect of violence, but a condition of violence – its effects intensify the impacts of violence, reinforce prevailing power relations, and deepen unequal outcomes between differently located victims and survivors. We noted that trauma is one of the mechanisms through which disasters become layered and compounded, and through which the effects of disaster continue to wind slowly on through time (see Pain 2021b; Schwab 2010). And the more layered and compounded disasters are, the more intense and long-lasting subsequent trauma may be. Similarly, Lohokare and Davar (2000) distinguish between 'episodic' and 'enduring' disasters, noting the differential impacts on mental health. Therefore, women who experience GBV after disasters are even more likely to experience serious effects of trauma than those experiencing only one or the other (First et al. 2017). Trauma, in turn, is compounded by the various effects of disasters and GBV, for example, impacts on income, livelihoods, environmental and cultural change. Trauma and its impacts are also closely related to place (Coddington and Micieli-Voutsinas 2017; Pain et al. 2020), including the availability of social and formal responses to it which may improve or worsen it.

Below, we consider some of the impacts of trauma and its relationship to GBV, disasters and place.

Psychological impacts of GBV and disasters

A marked finding from the Cumbria study (reported in Chapter 4) was the prevalence of traumatic symptoms among the female respondents. For most, these were connected to the flooding disaster alone, but for a few, they were exacerbated by experiencing domestic violence from partners at the same time or in the past; 33% of women said that the floods had made their health worse and the majority mentioned their mental health specifically. Stress, anxiety and depression caused by the floods were common, as was trauma that sometimes lasted years:

> *It worries me that all I had worked for over the years would be lost, and I find that I panic now if it rains hard for a long time. I have panic attacks for no reason.*
>
> *(Female, Cumbria)*

The study in Barguna (detailed in Chapter 2) showed that GBV during disasters has several types of psychological impacts on victim-survivors, many very similar to the symptoms that respondents in Cumbria report from the disaster. However, a significant difference is that none of the respondents in Bangladesh had sought formal diagnosis or treatment of their mental health problems. Yet our analysis of their reported symptoms after both disasters and GBV show strong similarities with psychological trauma, a feeling of 'intense fear, helplessness, loss of control and threat of annihilation' (Andreason 1985 cited in Herman 1997; 33), deep sadness and a loss of self-confidence. Specific symptoms of post-traumatic stress disorders such as hyperarousal and intrusion (Herman 1997) were also described by respondents.

Every warning period of cyclones brings back traumatic memories for some respondents. Trauma is intensified not only by the stress and damage caused by the cyclone, and the incidents of violence that were experienced, but also by seemingly more minor incidents of harassment and women's fears around the impacts that harassment can have on the culture they live in. They mentioned their fear of being molested in crowded cyclone shelters, which created hyperarousal for some:

> *I am afraid of the cyclone shelter. Last time I went there, someone pushed me so hard that I fell…I was pregnant that time…someone pulled my blouse and tore it.*
>
> *(Female, Barguna)*

However, this situation was more severe for women who were taking shelter without a male family member with them. This respondent mentioned her fear in the interview:

> *In the shelters…in the crowds…on the way to shelters some young men touch women, girls… they pull their clothes. If they see any beautiful girl or women sometime they pull them. I am alone, my husband left me…I am vulnerable too.*
>
> *(Female, Barguna)*

The experience of being harassed at the cyclone shelter does not stop there, rather it has huge impacts later on in women's lives:

> *I was in an embarrassing situation. I saved myself from the huge waves of storm surges but I lost my Sari. I was saved by a man and he took me to the shelter. I was so ashamed without my sari, wearing just a petticoat and blouse.*
>
> *(Female, Barguna)*

This woman's worries show that she could save herself from the huge waves, but she could not save her honour. She is ashamed that she lost her sari, and worried that an unknown man pulled her from the water, because this might impact the future of her marriage. As respondents mentioned, such incidents can have significant implications for their characters and reputations, which are of primary importance to their husbands, family and society:

> *During the warning time people run for their lives. There were many people... Muslims, Hindus run together to reach the safe shelters. My husband became suspicious, 'why were you running with that man?' I ran to save myself, I did not notice that... but he was not happy.*
>
> *(Female, Barguna)*

> *If bad men tried to touch a woman and her husband saw this, then she would be blamed for this. The man thinks his wife might be involved with that man.*
>
> *(Male, Barguna)*

This shows the complexity of the situations created by disasters for woman in a conservative gender-biased society, where women are always facing social tests they must pass to survive. It is not only cyclone impacts that we observed; women in Barguna are often facing socially created hazards along with 'natural' disasters.

In the Cumbria floods, more respondents mentioned their worries about property crime in the aftermath of the flooding than about being physically or sexually assaulted. Many had been concerned about leaving their flood-damaged properties, even as the water was rising, because of security of their property, and 8% reported that they had been burgled or been the victim of theft as a result of the floods:

> *Vans were coming from other areas and going into homes to take from the upper floors.*
>
> *(Female, Cumbria)*

> *Opportunist burglary was a problem with homes being empty for so long.*
>
> *(Female, Cumbria)*

A small number of the women had felt less safe from assault because of the quality of their temporary accommodation, and others reported harassment from the numerous workmen who came into the area to repair damage to housing:

> *There were difficulties locally for women dealing with contractors (all male).*
>
> *(Female, Cumbria)*

However, on the whole, the safety of female flood victim-survivors in Cumbria was not reported as being as severely compromised as it was for women in the Bangladesh studies.

Psychological impacts of domestic violence and its complexity

As we explained in Chapter 2, GBV victim-survivors in Barguna are in a trap created by gender identity, responsibilities, social attitudes, culture and tradition. They are abused in the name of tradition, their complaints are considered through the lens of tradition, and they often tolerate these forms of violence to maintain tradition and social status because of a lack of alternative courses of action. Most harmful of all is the inability to express their feelings because it is not socially acceptable. Besides, most of the time, if they shared their stories of abuse with others, they would be further humiliated, and so victim-survivors mostly try to keep their stories private. This significantly increases their vulnerability to mental health problems, as social support and sharing traumatic experience with others has an important protective effect (Herman 1997). Below, we summarise some of the psychological impacts of GBV which have been revealed from the fieldwork.

One respondent is a young woman, studying at university, who was married off at the age of 19. Her family arranged the marriage and she was happy with the proposal. However, after marriage, it became clear that the man was involved in adultery and he did not want to change this. He started avoiding her and then beating her. She tried her best to tolerate the abuse and made several attempts to continue the relationship and stay with her husband. But her husband continued to beat her for very trivial reasons, and one day it got worse so that she had to return to her father's house. Later, her husband abandoned her. She was very young when she married, and started out hoping for a happy life with her husband. However, her dream turned into nightmares as she describes, '*I was so shocked, sad and upset …there was not a single day that I could take my meal without tears*' (Female, Barguna). Within two years, she became a divorced woman, dealing with the physical injuries of domestic violence. Her social status has changed significantly within these two years as she has returned to her father's house.

Not all victim-survivors of domestic violence can return to their parents. The study reveals that most are relatively powerless to leave their husbands, with no option but to live in a violent environment and continue in a bad marriage, living in constant fear. Being beaten in one's own home does not only create mental pressure, immense fear, deep sadness and trauma, but the victim-survivor often loses a lot of self-confidence. These can be life-long mental health impacts. One victim-survivor of GBV mentioned in her interview that she could not work out why she had become victimised. She had lost her faith in life and in herself:

> *I do not know why my life has become like this. My husband left me just after 4–5 years of my marriage. I have to work at other people's house. People talk about me good and bad.*
>
> *(Female, Barguna)*

The study shows that those targeted by domestic violence are mostly young women who are first victimised at a very young age. They may be beaten and tortured by their husbands and in-laws. They describe passing many days of their married life in tears. Those who stay at their husband's houses remain in constant fear that he will become violent again, and are forced to abide by all his rules and orders to try to keep him calm:

> *I do whatever he says, never disobey him now...*
>
> (*Female, Barguna*)

> *I never ask for anything from him to buy, I do it myself, I want him to be calm.*
>
> (*Female, Barguna*)

One mother of two young girls has been beaten regularly by her husband from the beginning of her marriage. When her husband becomes violent, she hides in her neighbour's house for a day or night until her he calms down:

> *He beats me, evicts me out of my house at night. He even forced me to sit in water for hours. I have to hide at another house to keep me safe. My daughters provide me meals secretly. I cannot go to my parents house...I have to stay here like this.*
>
> (*Female, Barguna*)

Listening to her long story, it is clear that her only fault is that she is a woman. Many of the victim-survivors in the research live in a home-prison for their whole life without any other reason.

No place to go

Women who suffer domestic violence are mostly beaten and tortured for dowry money, for protesting the adultery of their husbands, or claiming their rights for food and shelter. Among these victims, many do not leave their abusive husbands. Instead, their husbands may leave them and remarry. The question arises why women who are regularly beaten by their husbands do not leave their marriage. The answer is complex. These women do not have any place to go. They do not have enough education to enable them to get a proper job and maintain their life. Most did not complete their primary education because of child marriage and poverty, so they cannot be economically self-dependent and receive social respect, but can only get low-paid jobs like housemaid or day labourer. However, working in informal jobs increases their insecurity to other forms of GBV like stalking, proposals to form illegal relationships, and rape threats, leading their social status to decline. According to the respondents, women do not want this alternative life, which is described as like jumping from the frying pan into the fire, and so many do not leave their husband's house. In Barguna culture, even returning to their father's house is the most shameful option for a woman. Besides, girls are married off to reduce the costs of their family, especially after disasters, so going back again does not solve the problem. So many married women try their best to stay with abusive husbands. Husbands may use these circumstances as an opportunity to threaten their wives, warning that they will be sent back to their parent's house if they protest against domestic violence:

> *If you go to your father's house now, then go for forever, don't dare to come back.*
>
> *(Male, Barguna)*

> *I will send you to your father's house and never bring you back in this life.*
>
> *(Husband of a domestic violence victim, Barguna)*

Women also think of their parents' honour, which would be affected by their daughter's broken marriage. Daughters' selfless tolerance and sacrifice in this regard are praised by society, and even by the parents. One mother of a GBV victim, who was describing her daughter's tolerance and capacity to keep their family's honour said with pride that her daughter says,

> *Let me see what happens, I do not want to be like her (one of her cousins who is divorced and living with parents), leaving my husband's home and become a problem for the family.*
>
> *(Female, Barguna)*

However, our research reveals that, nonetheless, this nightmare comes true for some women, who are forced to return to their parents. Many of them were abandoned by their husbands and very few took their own decision to leave the marriage. Either way, whether they leave or remain in their marriages, or their husband leaves, the mental health impacts are significant.

Abandonment results in great mental distress, as Kairun mentioned in her interview:

> *When a husband leaves a woman, she feels abandoned to herself and feels abandoned everywhere, even at her fathers' house.*
>
> *(Female, Barguna)*

Most victims who had to return to their parents feel very bad about it. They may be earning their own meals but still feel unwanted at their parents' house:

> *I do not know what should I do, I cannot go back to that house (in-laws' house), I cannot stay here (father's house) …he (husband) is enjoying his life, I saw him on Facebook…he went on a vacation.*
>
> *(Female, an educated young women who has to do home tuition to earn money, Barguna)*

Rima, a GBV victim-survivor, talked about the helplessness and vulnerability of women who have been abandoned by their husbands. Her experience also throws light upon the constant struggles with her disability and intense poverty, all of which contributed to her poor mental health and trauma. She is 45 years old, living in a one-room accommodation provided by the government for poor inhabitants after Cyclone Sidr. She has physical disability that has affected her walking since childhood after having severe typhoid fever, which also meant she had not completed a primary level of education. She was married early and has a daughter, but like many of the poorer women in Barguna, her husband cheated on her and remarried another woman. Rima works as housemaid and sews katha (a local blanket). Her earnings are very low. Her husband often comes, beats her and takes any help she receives as reliefs or help from the government or NGOs, and one day, he took her only goat to sell. He has threatened to murder her to buy a rickshaw van for himself. Her husband also spreads rumours in regards of her character. Rima did not file a divorce case or leave her husband, being afraid of losing her honour and the little security that being married affords her.

In some cases, women whose husbands do not take responsibility for them still live at the husband's house with their in-laws. However, these women are not welcome there, and face ill treatment and misbehaviour from their in-laws. Their husbands may have remarried, or left the house after cyclones and stopped contacting the family. In this situation, women experience great sadness, worry and tension. They use every opportunity and spend every penny to get news of their husband:

> *I asked everyone about my husband. Where he could go? No one could answer. I cried and cried for a long time.*
>
> *(Female, Barguna)*

These women live with the pain of betrayal:

> *I am not well, mentally. It looks like I am fine...actually I am suffering...my husband does not take responsibility, I struggle every day. I do not tell anyone...only a few close relatives know my situation.*
>
> *(Female, Barguna)*

Another effect of trauma among some victim-survivors is feeling that they are responsible for the GBV they have experienced. They feel abandoned, and take the blame for the violence from their in-laws. Just as in-laws are happy when they receive gifts (dowry) from the bride, victim-survivors understand it to be normal that parent in-laws become unhappy when they are a burden on the family:

> *My in-laws keep me with them after my husband left me. They sometime behave badly with me. They can do that as my husband does not take any responsibility, they have to feed an extra person.*
>
> *(Female, Barguna)*

Powerlessness and frustration

This feeling of powerlessness is reported by many respondents. When they do not get any help and support from the family, local authority and other institutions, they become frustrated. Shahida who lives with an abusive husband described her feelings as follows:

> *I rely on Allah now. If He wills everything would be fine again. I complained to the Chairman and Members. They said I have to be patient. I know I have to be here with my husband tolerating every torture. I do that. What can I do. I cannot go back to my parents, they are also very poor.*
>
> *(Female, Barguna)*

Moriyam explains her feelings after her husband left, experiencing deep frustration. She has stopped socialising and lives a conservative life:

> *I do not feel well...anytime. I cannot remember everything. After my husband left, there is no one close to me alive. My mother, father are dead. I do not feel any interest to visit the neighbours' house...I do my work, earn a small amount of money and just lie down on my bed silently... I tell my Creator please take me as early as possible... do not let me fall sick.*
>
> *(Female, Barguna)*

As these examples show vividly, it is not only incidents of violence which cause trauma and mental ill health effects. Where there is psychological and emotional abuse, manipulation, betrayal and mind games, and where these situations are ongoing through lack of ability to leave the abusive husband or in-laws, trauma deepens (Herman 1997; Pain et al. 2020). All this is not to say that our respondents do not practice resistance. Victim-survivors of disasters, violence and other forms of harm may be both traumatised *and* capable, always involved in self help and peer support, even in the most constraining and risky of circumstances (Akerkar and Fordham 2017; Pain 2021a). We turn to women's strategies for coping, mutual support and resistance in Chapter 7.

While self help and community/mutual support have always been the primary responses to trauma, and this is still the case in most places, formal responses to trauma and the other mental ill health conditions that often accompany it, such as depression and anxiety, also make a significant difference. As feminist psychiatrists such as Herman (1997) have argued, a therapeutic relationship which is based on trust and places the victim-survivor in control can be foundational in rebuilding and recovering from traumatic events. However, the availability of formal treatments via healthcare and psychological services is, of course, highly variable between different places. In a comparison of Bangladesh, the UK and Malaysia, we have previously noted markedly differential access to trauma therapy for victim-survivors of GBV, even in non-disaster times (Pain et al. 2020). In the

UK, such treatments are more widespread, but increasingly hard to access as cuts are made to public services under austerity, and private therapy is available only for the affluent. In Bangladesh, there is very little mental health provision available in public health services, and this varies across the country. In Malaysia, there is access for some victim-survivors, but mostly based on imported Western models of care that are not always culturally appropriate to local beliefs and experiences of GBV (see Pain et al. 2020; Sahdan et al. 2021). An inverse care law applies to trauma therapy in many places, where those who have had the most chronic traumatic experiences are least likely to be able to access care after disasters or violence. And during times of disaster, there is more strain on healthcare and other social services, more difficulty in physical and financial access for victim-survivors, and less time to do so (Rezwana 2018).

Trauma, then, is an important condition that is both exacerbated by and exacerbates the impacts of disasters and GBV, and is especially pronounced when both are experienced at the same time.

6.4 Gendering violence

Having considered some of the ways in which place affects levels and forms of GBV and trauma, we now move on to examine the pathways and processes by which certain people and groups become perpetrators, victim-survivors, or both. Our focus is on the ways in which violence becomes gendered. We note again that men are most likely to become perpetrators irrespective of geographical location. However, cases and evidence of women becoming abusers or supporters of domestic violence are also evident in the research reported in this book, and we extend our gendered analysis to these cases of GBV.

Pathways to becoming a perpetrator or victim: explanations and examples

A universal finding is that men are the main perpetrators and women are the main victim-survivors in GBV. According to Dawson (2017), domestic homicide is mostly committed against women by men, and this is very common globally. In Australia, one in six women has experienced physical or sexual abuse by a current or former partner, compared to one in 19 men (Dunkley and Phillips 2015 in Butler et al. 2017). In England and Wales, 46% of all female and 7% of all male homicide victims in 2013/2014 were killed by a partner or ex-partner, and while female victims were killed by men, male victims were killed by both male and female partners or ex-partners (Payton et al. 2017). In Scotland, 91% of those accused of homicide were men (The Scottish Government 2015). In Norway, 72.1% of male victims were exposed to violence from unknown persons, whereas female victims were most likely to have been exposed to violence by their partners (40.9%) (Thoresen and Hjemdal 2014 in Lien and Lorentzen 2019). In the USA, around three women a day are murdered by spouses or ex-spouses, there is a rape every 6.2 minutes and a beating every nine seconds; this is the main cause of death for pregnant women (Solnit 2014). According to Krahé (2018: 6), men's violent behaviour against women must be explained by addressing three levels – 'the macro level of society, the micro level of dyadic interactions and the individual level of perpetrator characteristics'.

The range of factors associated with femicide varies from country to country, but the basic cause of structural inequality reaches across the world (Dawson 2017). In the USA, the colonial institution of patriarchy and fear of losing power and control work to create the context for femicide. In Canada, a history of domestic violence, an actual or pending separation and obsessive behaviour by the perpetrator are the main triggers for murders of women. In the UK, people living unconventional lives or from chaotic homes, with poor school records, substance abuse problems or a history of criminal convictions are most at risk, within a cultural belief system where men expect the subordination of women. And in New Zealand, the loss of indigenous cultural knowledge and practices, along with disparities in basic needs among members of society and the normalisation of domestic violence in everyday life, work to create the context of femicide. One or two risk factors may work separately or in combination in some cases (Dawson 2017).

The context of GBV in lower-income countries seems distinct due to the greater presence of poverty at individual and societal levels. Poverty is recognized or mentioned as the main influencing factor of GBV by many victims/survivors, but of course, poverty does not work alone. Disaster-induced poverty along with poor prevailing economic conditions increases tension, stress and conflict between family members; yet reasons for domestic violence against women are more strongly and deeply rooted within societal systems. The causes are present long before disasters hit society. The following sections illustrate some of the contexts, reasons and factors that lead men and women to become perpetrators or victims, drawing on some of the research presented earlier in this book.

Men's controlling power and societal support: experiences from the field

Why do men so often become perpetrators and choose violence when experiencing the stress of disasters? In contrast, many women experience the same stresses – and the further stress of being victimised by their partners – but are not violent in response. Instead, they tend to keep violence secret, with most not reporting it (see Chapter 7). Solnit (2014) suggests that we understand violence against women better if we look at the abuse of power as whole, from the beginning of society and across people's lives. The role of GBV in controlling women has a long history. For example, until the Women's Property Acts of 1870 and 1882 in the UK, everything belonged to their husbands, even the wife herself, who was penniless and totally dependent on her husband, with her separate identity being erased after marriage (Solnit 2014). Patriarchy, the social system we introduced in Chapter 1 (see Box 1.3 in Chapter 1), was practised and 'confers greater levels of power and influence on males, [and] is the overarching factor that ties all of these mechanisms together' (Kilmartin 2015: 16). Today, this system is 'perpetuating unequal power sharing and relationships between men and women' (Mahapatro 2018: 15).

The research in Bangladesh shows that men, in conservative patriarchal societies, are viewed as a powerful creation of God in their upbringing, and are given a sense of their own power beyond their capacity, which sometimes helps them to overlook their limitations and their sense of humanity. These men may easily become violent and may beat their wives for dowry money, forget their promises, leave their children during disasters and get involved in extra-marital relationships. It is worth mentioning here that men's involvement in extra-marital affairs has a relation with economic conditions which is rarely found among women. Men are reported to get involved in extra-marital relationships when they earn less money compared to their wives, which threatens their masculinity. But also, men may also cheat when they are the main earner, to take advantage of their partner's economic dependency and inability to protest. Whereas only a very small percentage of economically dependent women cheat on their partners, and when women are the dominant earner in the family they rarely cheat either (Cole 2018).

Men from marginalized communities who have limited access to educational, occupational or economic success may use violence to achieve alternative forms of masculinity (Reilly et al. 2004). As Anwary (2015: 39) explains, 'gender subjectivity and foreclosure are formed discursively in the context of violence against wives in Bangladesh'. Men's subjectivity and its establishment in the family and society encourage them to control women and misuse their power – 'violence is one way to silence people, to deny their voice and their credibility, to assert your right to control over their right to exist' (Solnit 2014: 6).

However, most striking is that men's violent behaviour is often accepted by society (Parkinson and Zara 2013; Sloand et al. 2015), not only in Bangladesh but in other cultures too. For example, 73% of women in Solomon Island believe that man can beat his wife (UNWOMEN 2014). And in a study in Northern Ireland in the UK, young men expressed the view that young women respect violence in men and find it attractive, which points to wide social acceptance (Reilly et al. 2004). As Kilmartin (2015: 15) puts it, violence against women is 'partly embedded in the social meaning of what it means to be a man and in the social-structural conditions that create power imbalances between the sexes'. Applying a lifecourse perspective to GBV (see Mahapatro 2018), it can be seen that violence against women does not consist of separate incidents, rather it is an interlinked series starting from a woman's birth and sometimes even earlier. According to one of the respondents from Barguna,

> *Violence in Barguna includes rape, child marriage, wife battering, torture for dowry, mental torture. However, discrimination between girls and boys created from their birth by and within the family should be included in the definition of violence because it is the root of everything.*
>
> *(Male, Barguna)*

Discrimination between girls and boys was mentioned in several interviews, with one key respondent suggesting that

> *Most of the people believe that girls do not need higher education. There would be no profit for them if they educate daughters. Son will earn for them and take care of them. So, they should spend more money on sons.*
>
> *(Female, Barguna)*

Women as victims: power imbalances and self-confidence

At the same time, girls are raised in such a way that diminishes their sense of their capacity and their confidence. According to Anwary (2015: 42) in her paper on violence against wives in Bangladesh, 'women's independence and assertiveness seem to be a potential source for gender violence'. They feel themselves to be less than fully human, too dependent to protest. As Solnit (2014: 8) says of herself, 'most of my life, I would have doubted myself and backed down'; she describes how women fight on two fronts – one for whatever the topic is that she is talking about and the other for the right to speak. However, there is also a battle with herself to gain confidence to speak about either of these things. All these self-perceptions contribute to a huge power imbalance among the members of a society or family. Women and men behave, unintentionally or intentionally, according to the perceptions and confined gender roles society has fixed for them. These roles may be comfortable for men as they are on the privileged side. However, problems arise when some women protest against these gender roles, discrimination and violence.

> *He has beaten me until I became senseless. My mother was present at that time. She could not do anything. She took me home.*
>
> *(Female, Barguna)*

This respondent's quote shows the helplessness of a mother and a daughter in the face of violence, representing the experience of many Bangladeshi rural women in the research, who cannot do anything but cry on their own. Emotional abuse is absorbed within life seemingly so naturally that it becomes daily practice for powerful groups. While some women deal with low self-esteem for their whole life, gendered support from society helps some men to become oppressive and continue their behaviour. Men feel far more supported by their family and society from childhood; their privilege is deeply embedded (Parkinson and Zara 2013). They have little accountability to their family and even to society. These freedoms are often misused by men, for example, in rising rates of GBV in disasters.

Violence against women is sometimes legitimised in the name of tradition. As a male inhabitant in Barguna explains,

> *Our tradition was to think women were like subordinates. So men beat them. Still some men are not aware and they do the same...Actually, some women do not have a problem to be beaten like this. Men want to control and dominate women.*
>
> *(Male, Barguna)*

This male respondent finds it acceptable for men to beat their wives, and he feels that women do not mind it. A similar perception is mentioned by Mahapatro (2018), who explains how perpetrators use cultural rationalisations for violence against women in India (and see Sahdan et al. 2021 in Malaysia). Likewise, seeing elders misbehaving towards female relatives make this behaviour normal for people living in Barguna:

> *We saw our previous generations do the same with our mother, aunties and grandmothers and we just follow them. We were raised like this.*
>
> *(Male, Barguna)*

Similar justifications have been given by men living in other parts of the world. In New Zealand, the normalisation of domestic violence is revealed when men blame their experience of being brought up in a similar situation (Wilson 2017). Here, social learning theory helps to understand this phenomenon, as behaviours are passed through generations due to the absence of positive roles and any protest from the family and society (Hines and Malley-Morrison 2005 in Mahapatro (2018)). According to Mahapatro (2018), cultural norms around violence are not limited to the individual level, rather they are re-enforced by the family, community and the wider structures of the media and politics.

Men's violent behaviour in disasters: explanations from the field

Stress and tension in disasters and social acceptance

In the research presented in this book, a number of different reasons are given by respondents for the rise of GBV during disasters. As we saw in Chapter 4, however, in the Cumbria study, the overwhelming reason given by respondents (most of whom had not been victimised) for the rise in domestic violence was the stress associated with disasters. These stresses were experienced as multiple – around having to move out of their homes, the damage to housing and the long wait to get insurance money and repairs done, looking after children in cramped temporary accommodation, additional financial and health worries and the general disruption to people's lives. Many respondents thought that this led to arguments and conflict within the family, which might in some cases then lead to domestic violence. There was a clear divide noticeable between people who had personal experience of abuse themselves in the past – these respondents tended to believe that domestic abuse would be unlikely to happen for the first time during disasters, but would have been happening previously too – and those with no experience of it and who felt that domestic violence reflected couples' poorer skillsets and lower resilience in difficult situations: *'I think for some people it will, if people don't have coping skills and lack emotional support'* (Female, Cumbria).

These views reflect the fact that public understandings of domestic violence in the UK are quite limited, tending to put the blame on individual factors rather than seeing the occurrence of violence as a persistent and chronic problem that is underpinned by structural factors. As one respondent said, she would expect flooding to lead to increased domestic violence *'in the same way losing a football match apparently does – any hardship'* (Female, Cumbria).

In Bangladesh the idea of increased stress driving abuse was also a common suggestion. One of the male inhabitants said that if there were no cyclones in Barguna, there would be much less GBV. Some of the female residents claimed that their husband's behaviour started to change after disasters, as losing everything in disasters created great mental pressure and some engaged in negative activities to cope with this. One respondent's husband in Barguna became addicted to drugs due to those losses, and started committing GBV:

> *My husband became drug addicted after Cyclone Sidr. He lost everything in his shop and he started neglecting his responsibilities for his family. He became violent and beat me...Now I have to take the main responsibility for managing money for the children's education and other things.*
>
> *(Female, Barguna)*

Another female resident described the context of her abandonment after the disaster, and this trend is found to be common in the study area (see Chapter 2). According to her description, her husband's mental health problems started after Cyclone Sidr. Previously, he worked on a fishing boat and earned money that was enough for them to live on. However, after Sidr, he stopped working and started doubting his wife. He controlled her movements. He beat her if she wanted to go outside their home to earn their daily meal. Then after a few months, he left her. She did not know where he had gone.

In Cumbria, too, many respondents reported that their partners' mental health had deteriorated and that this impacted their relationship. These cases may help to explain some of the reasons and contexts for men becoming perpetrators due to the immense impacts of disasters. However, female respondents also express the depression, sorrow and stress they had felt after the disasters, as well as a range of effects on their mental health, but this rarely leads them to violence. Rather, they appear to tolerate an increased amount of violence without challenging it during the immediate post-disaster period, as they focus on the family's survival and often take on additional caring responsibilities. In Bangladesh, there was evidence that some were involved in forced physical relationships to earn relief and survive the difficulties of the post-disaster period.

This shows that there is a complex interrelationship between culture, gender roles and people's behaviour. In Bangladesh, it is socially expected and accepted that men can show their feelings publicly after disasters, whereas women should manage their feelings which remain private and hidden (Anwary 2015). Husbands may become drug addicted or flee from their family, but it is impossible to think of a Barguna woman doing the same due to the frustration caused by disaster impacts; or that she would leave her children, due to her heavy housework burden and responsibilities. A drug addict husband will still get shelter from his family and the society may turn a blind eye, but a woman would be evicted from her house and thrown into the street. In the UK, cultural expectations are that women will prioritise looking after their children no matter what. Therefore, sharp gender discrimination can be found in social attitudes, which also provide indirect support for men who become perpetrators. As respondents describe it in Bangladesh, a man does not have to think for a second before hitting his own wife, and there is support for men who conduct extra-marital relationships. One respondent explained,

> *When he (a perpetrator) married her, she was good looking. However, after few years due to poor diet and child bearing at the early age she lost her beauty which also became a reason for torture. Her husband started avoiding her and adultery started.*
>
> *(Male, Barguna)*

This is an important and alarming point from a male respondent who opened up to us and told us his opinion, which is to defend men in extra-marital relationships. If we are raising our male children to become perpetrators within this confined social system, GBV will remain high and may indeed increase in future.

Powerlessness

Another view expressed in the studies in Bangladesh is that men who are poor lose so much in disasters that it drives them to violence. Men lose their jobs, wealth and money in disaster and become helpless. However, when women are given priority after disasters rather than men, being given loans or relief by the disaster responders, which creates an imbalance in the prevailing power relations in the family and local community.

> *This policy makes men feel left out and sad. Actually men always want to be on the strong side.*
>
> *(Male, Barguna)*

His view reveals that some men expect women to be weak and dependent, and feel that getting relief from outsiders makes women 'more clever' and 'self-confident'. Besides, some men do not want to face the hardship of post-disaster periods for their families; so they try to escape. Some roam around and wait for what they see as easy money in the form of relief. This creates a crisis within relationships and increases domestic violence.

Controlling women and money

Another aspect that men blame in Bangladesh for GBV relates to their wives' behaviour. It was mentioned by several respondents that women often want to help their parents, especially after disasters, and so if women receive any relief or grants, they think of their parents as well as their own family. But this is not accepted in this patriarchal society, where the common practice is that women cannot take any more responsibility for their parents after marriage, even if they are earning. In Barguna, women who are employed and educated may be beaten and abused by their husbands in their own houses (see also Mahapatro 2018). They do not always get permission to help their parents from their earnings. In this way, men use violent acts to 'feminise' women and control the assertiveness which they gain from paid work (Anwary 2015). This culture is especially strongly practised in poor families where resources are limited. However, as a daughter, a woman desperately wants to help her parents, and sometimes takes the money from her husband without his permission. A male respondent mentioned in his interview that he knows that his wife does that all the time, but her habit became a mental torture for him during the post-disaster crisis period:

> *Some women take money from her husband's earnings and send to their parents secretly. After disasters this becomes a torture for the men.*
>
> *(Male, Barguna)*

Overall, then, respondents in Bangladesh had more familiarity with domestic violence and blamed more complex and multiple factors for its raised incidence after disasters. In the UK, where respondents generally had less familiarity with it, it was predominantly blamed on the stress of disasters. In Bangladesh, some respondents blamed the conservative patriarchal society directly. On the other hand, some accounts tended to excuse, or even in a few cases, support men who abuse their wives.

When women abuse men and other women, and men are victims

Power dynamics and GBV

So far, this chapter has shown that the context, reasons and influencing factors of GBV are often multiple, with complex and layered power dynamics involved in its causes. Rather than power being something that is attached to gender in a rigid and immovable way, anyone (male or female) who has power (economic, social or political) may abuse it to become involved in violence. In particular, with GBV, 'the family is an institution of power and each of its members applies some degree of power', and 'those with access to more resources have more power' (Anwary 2015: 39). This understanding of power as a flexible rather than fixed resource means that women are sometimes violent towards men and to other women, depending on their own relationship with different forms of power.

In Bangladesh, when women belong to wealthy families and bring valuable gifts from their parents' homes to their husbands', in some cases, they go on to behave harshly towards their husbands and their family. Men mention these behaviours as psychological violence against them in some studies of cyclone- and flood-prone regions (Lien and Lorentzen 2019). In a few cases, women were involved in extra-marital relationships which men mentioned as socially humiliating and causing them mental torture. The number of conflicts between husband and wife increases during the post-disaster periods, when women are reported as participating in quarrels and 'situational couple violence' (Lombard 2013). The major reasons given for these incidents are men's inability or unwillingness to earn money during emergency periods which leads to children going hungry, women staying at home, and a lack of money for healthcare and for repairing damaged houses (Stark 2013). In addition, men may be using drugs or alcohol, which is perceived as wasting the family's or woman's money. These women become violent towards men to protect themselves and their children from alternative physical, mental and economic hardships. In these cases, violence is a defence mechanism for women, similar to those found among some women involved in IPV in the USA and India (Mahapatro 2018; Websdale et al. 2017).

However, rates of domestic violence against men remain very low compared to GBV committed against women. Lien and Lorentzen's (2019) research revealed that out of 700 male respondents, only 9% reported violence against them by the people with whom they have an intimate relationship. Moreover, women's violent behaviours tend to differ qualitatively, as there were no injuries reported in these studies, rather verbal and psychological abuse. More broadly, 'men and women use and experience violence differently and the impact of violence is different...Men who experience violence are less likely to live in fear' (Lombard 2013: 178) and violence is committed by women with less desire to control their partners. Whereas violence against women is more likely to be serious or deadly, to occur repeatedly, and to involve coercive control as one of the main aims.

A few GBV incidents were revealed in the Bangladesh studies (Chapters 2 and 3) where men became victims of sexual harassment or abuse, especially during their childhood. In these cases, men were comparatively weaker due to age and socio-economic status, a relation between power dynamics and GBV that is common all over the world. Recently, news of GBV in the workplace in higher-income countries has come to attention all over the world. Along with women, some men also came forward with experiences of sexual harassment, often in a powerless position compared to their bosses. They described that they were in too weak a position in the hierarchy to protest, and in fear of losing their jobs. Their male bodies and identities did not help them in these circumstances, as ultimately GBV is about forms of power beyond the physical. It is important to recognise, too, that these gender identities and power relations sometimes work negatively for men. They do not share their experiences with others, preferring to conceal those incidents (Lien and Lorentzen 2019). Men in our Bangladesh research, too, reported that due to the gendered expectations of society they cannot talk to anyone about the violence committed against them, especially child abuse or verbal or psychological abuse by wives. They feel too ashamed and afraid of humiliation and that this would suggest weakness. Similar findings were revealed among Norwegian men staying at a refuge (Danielsen 2013 in Lien and Lorentzen 2019; Grøvdal and Jonassen 2015). Thus, GBV against men goes unnoticed, unrecognized and is largely kept secret.

However, childhood sexual and emotional experiences, including abuse, have a strong relation with men's violent behaviour towards their partners during their adulthood (Fulu et al. 2013).

When women abuse women

Women sometimes take part in domestic violence alongside men. We described earlier how mothers-in-law (who hold a powerful role in the family in Bangladesh) are sometimes found to be directly involved in dowry violence and mental torture. Here, age and assigned roles in the family work to bestow power and lead women to behave as perpetrators. Often being oppressed themselves at earlier stages of their life, these older women begin to abuse younger women when they achieve a more powerful role.

However, men are the main perpetrators in most GBV. As we have described, in the patriarchal conservative society of Bangladesh, men are raised with special attention and given extra power to be seen as superior in the family. Society often supports men in every aspect, even if they become violent and commit crimes such as wife beating, dowry, rape and eve teasing. This gendered social behaviour creates sharp imbalances, alerting us to the fact that so long as children (male or female) are raised with differential privileges and gendered societal support, it will be very difficult to reduce the rate of GBV. Disasters and GBV will remain as a double burden for women in disaster-prone regions, remaining a key under-recognised issue for disaster managers.

6.5 Conclusion

This chapter has drawn on a number of studies and the wider literature in different parts of the world, to analyse some of the relationships between disasters, GBV, culture and place. While the correlation between disasters and GBV has been demonstrated around the world, shown particularly during the Covid-19 pandemic, our aim in this book has been to try to understand *why* this connection is so widespread and so enduring across very different contexts. Our studies have shown that the reasons and triggers causing GBV during disasters are various and quite wide-ranging and that there are similarities and differences between countries such as the UK and Bangladesh, but also within countries such as Bangladesh. This in turn relates to variations in aspects of place and in local cultures. While we have argued that all patriarchal societies around the world create the conditions for violence to increase during disasters, there are variations in terms of who is involved, the forms of GBV, differences according to age and socio-economic status, and the responses of victim-survivors that are possible in view of the informal and formal support that is available to them. Places, we have argued, are affected by shifting layers of structural violence, caused by forms of capitalism, racism and colonisation as well as patriarchy, and these frame experiences of GBV in each place.

We have also shown how the experience of both disasters and GBV becomes compounded by trauma and other mental health effects. These were commonplace in the UK and the Bangladesh studies, and sometimes affected victim-survivors for many years. Where people have fewer social and financial resources, and there is less access to formal support, this trauma is likely to affect them for longer, continuing to cause material and social damage just as the initial disaster or incidents of GBV did. Of course, both GBV and disasters tend to be chronic, repeated experiences rather than one-offs, which make any sense of recovery of physical and mental wellbeing more elusive particularly when they are experienced together as a layered disaster.

Finally, this chapter has examined some of the reasons why men, and sometimes women, commit GBV. While many respondents in both the UK and Bangladesh blame the rise of GBV on the stress associated with disasters, there are fewer explanations given as to why this stress leads far fewer women to perpetrate abuse. We have shown that is the distribution of power and general attitudes to women in patriarchal societies around the world that underlie GBV, whether in disasters or at other times. This point has important consequences for the next and final chapter (Chapter 7), where we move on to look at responses to addressing GBV in disasters.

References

Abramowitz, S. & Moran, M. H. (2012). "International human rights, gender-based violence, and local discourses of abuse in postconflict Liberia: a problem of 'culture'?" *African Studies Review* 55(2): 119–146.

Akerkar, S. & Fordham, M. (2017). "Gender, place and mental health recovery in disasters: addressing issues of equality and difference." *International Journal of Disaster Risk Reduction* 23:218–230.

Andreason, N. C. (1985) "Posttraumatic stress disorder." In Kaplan, H. I. & Sadock, B. J. (eds.) *Comprehensive Textbook of Psychiatry*. Baltimore, Williams and Wilkins: 918–24.

Anwary, A. (2015). "Construction of hegemonic masculinity: violence against wives in Bangladesh." *Women's Studies International Forum* 50: 37–46.

Bradshaw, S. & Fordham, M. (2015). "Double disaster: disaster through a gender lens." In Shroder, J. F., Collins, A. E., Jones, S., Manyena, B., & Jayawickrama, J. (eds.) *Hazards, Risks and Disasters in Society*. Cambridge, USA, Academic Press: 233–251.

Butler, A., Buxton-Namisnyk, E., Beattie, S., Bugeja, L., Ehrat, H., Henderson, E, & Lamb, A. (2017). "Australia." In Dawson, M. (ed.) *Domestic Homicide and Death Review: An International Perspective*. London, UK, Palgrave Macmillan: 125–158.

Chanthy, S. & Samchan, H. (2014). *Flood Impacts on Women: Exploring the Possibility of Gender Sensitive DRRplanning*. Actionaid, Cambodia. https://reliefweb.int/sites/reliefweb.int/files/resources/KHM_0601_ImpactonWomen_AA2014-01.pdf, accessed on 18/07/2021.

Coddington, K. & Micieli-Voutsinas, J. (2017). "On trauma, geography, and mobility: towards geographies of trauma." *Emotion, Space and Society* 24: 1–5.

Cole, N. L. (2018). "Sociology Explains Why Some People Cheat on Their Spouses." http://thoughtco.com/why-people-cheat-on-their-partners-3026688, accessed on 26/08/2021.

Danielsen, E. M. (2013). 'Jeg Følte Jeg Var En Slags Merkelig Fugl Som Hadde Flydd Inn Her": En Kvalitativ Intervjustudie Av Menns Møte Med Krisesentre ("I Felt Like a

Strange Kind of Bird Who Had Flown in Here": A Qualitative Interview Study of Men's Encounter with Crisis Centres). Master's dissertation, Department of Sociology and Social Geography, University of Oslo. https://www.duo.uio.no/handle/10852/36850, accessed on 08/04/22.

Dawson, M. (eds.) (2017). *Domestic Homicide and Death Review: An International Perspective.* London, UK, Palgrave Macmillan.

Decker, M. R., Latimore, A. D., Yasutake, S., Haviland, M., Ahmed, S., Blum, R. W., Sonenstein, F., & Astone, N. M. (2015). "Gender-based violence against adolescent and young adult women in low-and middle-income countries." *Journal of Adolescent Health* 56(2): 188–196.

Dunkley, A. & Phillips, J. 2015. Domestic Violence in Australia: A Quick Guide to the Issues, Research Paper Series 2014–2015, Department of Parliamentary Services, Canberra.

Dustin, M. (2016). "Culture or masculinity? Understanding gender-based violence in the UK." *Journal of Poverty and Social Justice* 24(1): 51–62. doi:10.1332/175982716X14525979706964

Dutta, P. M. (2017). "Natural disaster and vulnerability to trafficking of women and girls in India." *European Scientific Journal* 7881: 54–69.

Enarson, E., Fothergill, A., & Peek, L. (2018). "Gender and disaster: foundations and directions for research and practice." In Rodriguez, H., Donner, W., & Trainor, J. E. (eds.) *Handbook of Disaster Research.* New York, Springer: 205–223.

First, J. M., First, N. L., & Houston, J. B. (2017). "Intimate partner violence and disasters: a framework for empowering women experiencing violence in disaster settings." *Affilia* 32(3): 390–403.

Fluri, J. L. & Piedalue, A. (2017). "Embodying violence: critical geographies of gender, race, and culture." *Gender, Place & Culture* 24(4): 534–544. doi:10.1080/0966369X.2017.1329185

Fulu, E., Jewkes, R., Roselli, T., & Garcia-Moreno, C. (2013). "Prevalence of and factors associated with male perpetration of intimate partner violence: findings from the UN Multi-country Cross-sectional Study on Men and Violence in Asia and the Pacific." *The Lancet Global Health* 1(4): e187–e207.

Grøvdal, Y., Jonassen, W., & Nasjonalt kunnskapssenter om vold og traumatisk stress (The Norwegian Centre for Violence and Traumatic Stress Studies). (2015). *Menn på krisesenter* (Vol. 5/2015, Rapport (Nasjonalt kunnskapssenterom vold og traumatisk stress: trykt utg)). Oslo: Nasjonalt kunnskapssenter om vold og traumatisk stress (The Norwegian Centre for Violence and Traumatic Stress Studies). Retrieved from https://www.nkvts.no/rapport/menn-pa-krisesenter/

Herman, J. L. (1997, first edition) *Trauma and Recovery.* New York, Basic Books.

Hines, D. A. & Malley-Morrison, K. (2005). *Family Violence in the United States: Defining, Understanding, and Combating Abuse.* Thousand Oaks, USA, Sage Publications.

Horii, H. (2020). "Child marriage as a 'solution' to modern youth in Bali." *Progress in Development Studies* 20(4): 282–295.

Kaufman, M. (1987). "The construction of masculinity and the triad of men's violence." In Kimmel, M. S. & Messner, M. A.. (eds.) *Beyond Patriarchy: Essays an Pleasure, Power and Change.* Toronto, USA, Oxford University Press: 13–25.

Kilmartin, C. (2015). "Men's violence against women: an overview." In Johnson, A. J. (ed.) *Religion and Men's Violence Against Women.* New York, Springer: 15–25.

Krahé, B. (2018). "Violence against women." *Current Opinion in Psychology* 19: 6–10.

Lien, M. I. & Lorentzen, J. (2019). *Men's Experiences of Violence in Intimate Relationships.* Switzerland, Palgrave Macmillan.

Lohokare, M. & Davar, B. V. (2000). "Women in disasters and mental health." *Indian Journal of Social Work* 61: 565–580.

Lombard, N. (2013). "What about the men? Understanding men's experiences of domestic abuse within a gender-based model of violence." In Lombard, N. & McMillan, L. (eds.) *Violence Against Women: Current Theory and Practice in Domestic Abuse, Sexual Violence and Exploitation*. London and Philadelphia, Jessica Kingsley Publishers: 177–193.

Luft, R. E. (2016). "Racialized disaster patriarchy: an intersectional model for understanding disaster ten years after Hurricane Katrina." *Feminist Formations* 28(2): 1–26.

Mahapatro, M. (2018). *Domestic Violence and Healthcare in India: Policy and Practice*. Singapore, Springer.

Montesanti, S. R. (2015). "The role of structural and interpersonal violence in the lives of women: a conceptual shift in prevention of gender-based violence." *BMC Women's Health, Springer Link* 15(1): 1–3.

O'Brien, J. E. & Macy, R. J. (2016). "Culturally specific interventions for female survivors of gender-based violence." *Aggression and Violent Behavior* 31: 48–60.

Pain, R. (2021a). "Collective trauma? Isolating and commoning gender-based violence." *Gender Place and Culture* doi:10.1080/0966369X.2021.1975103

Pain, R. (2021b). "Geotrauma: violence, place and repossession." *Progress in Human Geography* 45(5): 972–989.

Pain, R., Rezwana, N., & Sahdan, L. (2020). "Trauma, gender and space". In Datta, A., Hopkins, P., Johnston, L., Olson, E., & Silva, J.M. (eds.) *Routledge Handbook of Gender and Feminist Geographies*. London, Routledge: 287–296.

Parkinson, D. & Zara, C. (2013). "The hidden disaster: domestic violence in the aftermath of natural disaster". *The Australian Journal of Emergency Management*. 28(2): 28–35.

Payton, J., Robinson, A., & Brookman, F. (2017). "United Kingdom". In Dawson, M. (eds). *Domestic Homicide and Death Review: An International Perspective*. London, Palgrave Macmillan: 91–124.

Pruitt, L. R. (2008). "Place matters: Domestic violence and rural difference." *Wisconsin Journal of Law, Gender and Society* 23: 347.

Reilly, J., Muldoon, O. T., & Byrne, C. (2004). "Young men as victims and perpetrators of violence in Northern Ireland: a qualitative analysis." *Journal of Social Issues* 60(3): 469–484.

Rezwana, N. (2018). *Disasters, Gender and Access to Healthcare: Women in Coastal Bangladesh*. Oxon, UK, Routledge, Taylor & Francis Groups.

Sahdan, Z., Pain, R., & McEwan, C. (2021). "Demonic possession: narratives of domestic abuse and trauma in Malaysia". *Transactions of the Institute of British Geographers*. doi:10.1111/tran.12485

Schwab, G. (2010) *Haunting Legacies: Violent Histories and Transgenerational Trauma*. New York, Columbia University Press.

Seager, J. (2014). "Disasters are gendered: what's new?" In Singh, A. & Zommers, Z. (eds.) *Reducing Disaster: Early Warning Systems for Climate Change*. Dordrecht, Netherlands, Springer: 265–281.

Sloand, E., Killion, C., Gary, F. A., Dennis, B., Glass, N., Hassan, M., Campbell, D. W., & Callwood, G. B. (2015). "Barriers and facilitators to engaging communities in gender-based violence prevention following a natural disaster." *Journal of Health Care for the Poor and Underserved* 26(4): 1377–1390.

Smith, K., Osborne, S., Lau, I., & Britton, A. (2012) *Homicides, Firearm Offences and Intimate Violence 2010/11: Supplementary Volume 2 to Crime in England and Wales 2010/11*. London, Home Office.

Solnit, R. (2014). *Men Explain Things to Me*. London, Granta Publication.

Stark, E. (2013). "Coercive control". In Lombard, N. & McMillan, L. (eds.) *Violence Against Women: Current Theory and Practice in Domestic Abuse, Sexual Violence and Exploitation.* London and Philadelphia, Jessica Kingsley Publishers: 17–33.

The Scottish Government (2015). Homicide in Scotland, 2014-15. https://www.gov.scot/publications/homicide-scotland-2014-15/, accessed on 25/07/2021.

Thoresen, S. & Hjemdal, O. K. (2014). *Vold og voldtekt i Norge: En nasjonal forekomststudie av vold i et livsløpsperspektiv (Violence and Rape in Norway: A National Prevalence Study of Violence in a Lifespan Perspective)* (Vol. 1/2014, Report (Nasjonalt kunnskapssenter om vold og traumatisk stress: publication) (The Norwegian Centre for Violence and Traumatic Stress Studies)). Oslo: Nasjonalt kunnskapssenter om vold og traumatisk stress. (The Norwegian Centre for Violence and Traumatic Stress Studies).

UNICEF (2021). Ending child marriage. https://www.unicef.org/bangladesh/en/ending-child-marriage, accessed on 15/12/2021.

UNWOMEN (2014). Climate change, disasters and gender-based violence in the pacific. www.asiapacific.unwomen.org, accessed on 20/03/2021.

Websdale, N., Celaya, A., & Mayer, S. (2017). "United States". In Dawson, M. (ed.) *Domestic Homicide and Death Review: An International Perspective.* London, UK, Palgrave Macmillan: 27–58.

Wild, J. L. (2020) "Domestic Violence and Abuse: Prevention, Intervention and the Politics of Gender." PhD Thesis, Leeds University.

Wilson, D. (2017). "Indigenous populations and the domestic violence death review process". In Dawson, M. (ed.) *Domestic Homicide and Death Review: An International Perspective.* London, UK, Palgrave Macmillan: 287–316.

World Health Organization (2021). *Factsheet: Violence Against Women.* https://www.who.int/news-room/fact-sheets/detail/violence-against-women, accessed 05/04/22.

7

RESPONDING TO GENDER-BASED VIOLENCE DURING DISASTERS

Nahid Rezwana and Rachel Pain

7.1 Introduction

In this final chapter, we turn to disaster management and recovery. The book's aim has been to interrogate the relationship between disasters and gender-based violence (GBV), which together create layered disasters. The book has emphasised, firstly, how people's vulnerability to this layered disaster, and especially the vulnerability of women and girls, is primed by chronic structural conditions before disasters occur. Secondly, the chapters have shown how vulnerability is created during disasters by a range of practical and cultural issues that particularly affect women in different locations. Thirdly, the empirical studies yielded numerous examples of the ways in which the after-effects of disasters further compound vulnerability to GBV, through impacts such as ill health, the loss of employment or increased poverty, so that the trauma of layered disasters becomes long-term and is often repeated over time. These three interrelated dimensions – of before, during and after disasters – work together to create violence and unsafety, compounding layered disasters.

We have said less so far about a fourth dimension, that of disaster responses, yet these responses are also important in framing disasters. For example, some commentators argue that the disaster responses of governments and NGOs may sometimes worsen the impacts of disasters, or entrench rather than reduce vulnerability, making further disasters more likely (Appiah 2021; Jacobs 2019, 2021; Kelman 2020; Luft 2016; Mileti 1999). Such critiques have been made strongly by feminist writers, many of whom contend that gender, and particularly the structural analysis of gender inequalities that has been central to our approach in this book, is too often overlooked or the subject of responses that are superficial and therefore inadequate (Enarson et al. 2018; Ensler 2021; Seager 2014).

However, as we pointed out in Chapter 1, there is a whole range of actors who respond to disasters and the increases of GBV that are always associated with

DOI: 10.4324/9781003089780-7

them. In many ways, the 'first responders' are victim-survivors themselves. While conventional analysis of disaster responses is only concerned with the formal agents who partake in 'disaster planning', 'disaster management', and 'disaster recovery' – in other words, national and local governments, NGOs, and expanding private sector provision – it is vital to recognise the role of GBV victim-survivors and their communities in the work of response and recovery. This chapter therefore focuses both on existing policy and practice around disasters, and on the actions that victim-survivors take to save themselves and their loved ones and rebuild their lives, livelihoods and communities in the long term. They also take action to mitigate the risk and impacts of future disasters: victim-survivors are disaster managers and this is central to feminist and anti-racist perspectives. As we argue towards the end of this chapter, this work must be recognised and supported, and inform the efforts of formal agencies if these are to be more successful in tackling GBV during disasters.

This chapter, then, focuses on the implications of the research and arguments presented in this book for disaster planning, management and recovery. The first section begins with victim-survivors: drawing on examples from our case studies that bring to light their various strategies of coping and resistance when experiencing GBV and disasters. We then assess the limits of formal responses for these layered disasters, drawing on a range of feminist and related critiques. The following section then identifies our recommendations for policy and practice, which are informed by these literatures but also, importantly, by what survivors have told us in the field studies presented in this book. Finally, we conclude by returning to our conceptual framework of layered disasters. We reflect on the fit of what we have learnt about disaster responses in this chapter with Luft's (2016) notion of racialised disaster patriarchy, and we highlight the positive role of feminist and other social movement responses in communities affected by the Covid-19 pandemic, arguing that this is key to surviving and challenging politically and socially constructed layered disasters.

7.2 Victim-survivors' strategies: growing resilience or living as victims?

As victim-survivors are always the first responders to disasters, we begin with their accounts of coping, survival, resistance, and recovery. Of course, disasters and their magnitude and impacts vary widely, and so does the agency that victim-survivors are able to exert and the support they receive from each other and from the communities around them. Moreover, as we have detailed, survival is not an equitable endeavour but one that is cross-cut by gender, race, ethnicity, class, disability, age, sexuality, religion, caste, livelihood and migrant status (Sultana 2021). Nonetheless, as feminist writers have insisted, agency and resistance exist in the most constrained and risky circumstances of violence and abuse (Pain 2014). With domestic abuse in particular, it is often assumed that victims are partly responsible for their failure to challenge their partner's violence, but in reality, their resistance involves many different strategies that are often unseen (Enadner and Holmberg 2008). Resistance in the shape of direct oppositional action that challenges oppressive practices and structures may not be possible; it may simply involve practices that enable survival without changing the conditions of power (Katz 2008). In domestic situations of danger, it is often 'small acts' that constitute resistance (Pain 2014).

In the context of gender and disasters, Enarson et al. (2018: 205) suggest that just as gender shapes disaster vulnerabilities, it is vital to examine 'how agency and resilience are realized in crises'. This survival work on the part of victim-survivors is, we would argue, constant and ongoing, and central to what has widely been understood as disaster *recovery* by other people, organisations or states (see Box 1.3 in Chapter 1). We understand survival to be a long-term endeavour as well as an immediate response to an impending disaster; and recovery is rarely a return to previous life or to 'normality'. For many of the women in our research, disasters were not abnormal experiences but, especially for the older women, they are repeated, expected, and accommodated as part of life. Recovery work is undertaken, often tentatively, precariously and with fewer resources, and sometimes while repeated layers of disaster are being experienced. And responses to disasters, recovery and survival are strongly influenced by culture and place (Chapter 6) and by the availability of informal and formal support.

The case studies of research presented in this book reveal that survival strategies to cope with GBV during disasters are very different for men and women, as their experiences of disasters and disaster impacts usually differ. Men face GBV at a much lower rate during and after disasters, whereas women experience different forms and layers of GBV to an alarming degree (Rezwana and Pain 2020). Our research asked the GBV victim-survivors who participated in the research what coping strategies they use to deal with GBV. Do they protest the many forms of violence described in this book (Chapters 2–5), and do they develop resilience, or do they withstand violence, living as victims for their whole lives? The findings from the Bangladesh studies reveal that men usually take control of the situation when they face any confrontation or abuse. Most of the conflicts reported between husband and wife end with verbal abuse of women or other forms of domestic violence. Anwary's (2015) research explains that some men become violent, beating and even murdering their wives, when women are perceived to transgress their control and threaten their masculinity. However, our analysis shows that most female victim-survivors often do not want to let anyone outside the family know about the violence until it becomes very severe. Some do not even let their parents know about their suffering. Nonetheless, women also take a range of measures to cope with and reduce GBV, as follows.

'I tolerate everything'

The most common strategy that women in the Bangladesh studies apply when they encounter GBV during or after disasters is to remain patient and adjust. They often try to tolerate their husbands' abusive behaviour, initially at least, remaining patient in the face of lack of food and medicine, and managing with limited resources without complaining about the negligence of their husbands. As Lien and Lorentzen (2019: 57) remark, 'eroded self-esteem and fear of doing and/or saying something wrong are common consequences of living with conflict and violence', and so victim-survivors often do not show their true feelings to the perpetrator, as a way to keep them calm. Shahida explained the situation in her own words:

> *I am a woman, what else I can do? I tried to keep my composure. I know if I ask anything, his impatience or arrogance will be upon me, he will be violent. I tried my best to ask for less as much as I can. As a woman I do not have any choice but keep patient and tolerate everything.*
>
> *(Female, Rangpur)*

Nasreen's (2008) study also revealed this strategy was taken by female victim-survivors of domestic violence during floods.

When victim-survivors were asked why they do not talk about or report domestic violence, they mentioned several reasons behind this decision. The main reason is being frightened to lose their only shelter, their husband's house. We found that women and children become highly vulnerable when they leave or are evicted from their husband's home (see Chapter 2), and so women know that they have to live there until their husband evicts or divorces them. If they complain, the violence might increase and the divorce process may be accelerated, which often makes them more vulnerable still:

> *I do not complain to anyone because I know that I have to live in this house again. I have nowhere to go. If I tell anyone the torture might increase. So, I do not tell anyone...I tolerate everything.*
>
> *(Female, Barguna)*

> *He will be more arrogant if I complain and file a [legal] case. The Chairman [of the village] might say something or beat him, but then he will beat me again. Besides, where I will get money for the case?*
>
> *(Female, Barguna)*

These are rational decisions to remain, made with the family welfare and wellbeing in mind. Besides this, thinking of children's education and future opportunities is another important reason for victim-survivors to continue to tolerate domestic violence. In these regions, they know that divorce will bring more suffering for their children. Women know that they lack the power to make a stand against decisions taken by their husbands if they are uneducated, have no money, come from poor families and especially if they have been married off at an early age to reduce the family's responsibility. They also know about the stigma attached to women who are divorced or separated from their husbands (Anwary 2015). Many feel they will not get any help from their family, from the government or any other organisation. According to Anwary (2015), social expectations of violence against women and their economic dependency on their husbands both lead to justifications of violence against women as natural; and 'these forces convince abused women of Bangladesh not to leave their abusive homes in grave situations' (Anwary 2015: 41).

In instances where mothers divorce and remarry, they often have to leave their children, as new marriages do not allow children from a previous relationship. This is common practice in the study areas. Women do not want to leave their children, but on the other hand, they know that if they do not get married and try to survive on their own, it will often be impossible for them to provide a safe shelter, education for children and arrange good marriages for their daughters without their husband's support. Social humiliation is also a very important factor. One respondent described the situation like this:

> *I did not complain to anyone as people will say, you have got only a divorce by filing the case. It will bring dishonour to me. I will be more humiliated. Many women do not tell anyone only for this reason.*
>
> *(Female, Barguna)*

Parents of victims of domestic violence often also advise them to tolerate the torture for their family's honour and for their children's future. A similar situation was found in Australia after the 2009 Black Saturday bushfires. Parkinson and Zara (2013: 31) discuss how women were silenced around their experiences of domestic violence, reporting that 'there was enormous pressure from family members, friends, police and even health professionals to deny or forgive men's violence'. This is particularly pressing when a disaster has happened: women are expected to put others' needs first. Victim-survivors in our studies in Rangpur and Barguna in Bangladesh also try to avoid sharing their experiences with anyone because they may not be interested in listening. According to one respondent, '*People sometimes say, it is your family matter [wife beating and abandonment]... what we can do? So I do not tell anyone*' (Female, Rangpur). However, some women had shared their experiences for the first time with a friend who might also be a victim-survivor. They describe crying together and trying their best to keep their bad marriages going. '*Her husband also beats her. She comes to my house. We share our experiences. We cry together*' (Female, Barguna).

Some women also expressed hope that their husbands might become good again, and love them and take care of them, perhaps when a period of time has elapsed after the disaster. It would be ideal if their wishes came true, but no such cases were reported.

Taking financial responsibility

Some women take their own measures to cope with the economic crisis as well as to reduce their vulnerability to domestic violence after disasters. For example, going to stay at their parents' house during the floods may reduce the daily expenses of the family, avoid various problems created by the floods, and protect them from domestic violence. Nasima, in Rangpur, did this during a recent flood when her husband stopped earning. There was no food at home, but her husband did not take the initiative to get more food or money. Nasima was seven months pregnant and she and her unborn baby could not live without adequate food, so she requested her father to take her back home. They started walking in floodwater as the roads were submerged and there was no vehicle available. However, she was injured by floating objects which hit her while she was walking through the water, and her child was stillborn as a result. Nasima still remembers that flood with great sadness and talks of her helplessness and lack of choices as a woman. She had to return to her husband after the flood had subsided.

Women also get involved in earning money to recover from the economic crisis of the post-disaster period, though their contributions rarely get recognition as a part of the family income. Their goal in so doing is to reduce the impact of poverty, as well as protect against further domestic violence which they describe as increasing when money is short. So some start working outside the home after the floods, as labourers in the crop fields or the brickfields. They may migrate to other cities to work as housemaids or in garment factories. Leaving children at home is the hardest part of a woman's decision to migrate. Besides, 'such migration rarely ensures women's position in the family' (Hoque and Kamruzzaman 2016: v), and their reputation is easily tarnished when they are forced to migrate alone. Still, a number of women of Kashiganj village, Rangpur were found to have migrated to earn money to cope with the impacts of the flooding and the GBV that followed it. Research suggests that women never neglect their family's needs because they are not valued; on the contrary, they always seek ways to cope with post-disaster economic crises (Enarson and Chakrabarti 2009).

Another strategy, for some poorer women, stays hidden within the family; women may become involved in prostitution as a way of coping with disasters and GBV. This is a very sensitive issue and it was a big challenge for our research to collect information on it (see Chapter 3). It is mostly women from very poor families who get involved in this occupation, using their home as their workplace. Other women get involved in extra-marital relationships to get support for their subsistence when their husbands cannot earn enough to support them and their children.

Escaping and resisting violence

Because of the very difficult social and economic constraints discussed above, only a few of the victim-survivors in the Bangladesh research studies had taken steps to leave their abusive partner in an attempt to escape the violence. One reported in her interview:

> *If my husband loved me and all other members of his family did not like me and abused me, still I would have stayed with him...but if he does not like me, he beats me often, then how can I stay there? I left that house and took my children with me. I became the poorest and I have to struggle to earn money. I am still living on my own.*
>
> *(Female, Barguna)*

This woman wanted to continue her relationship but felt she had to leave to maintain her self-respect. She chose a life of struggle to live with what she saw as dignity.

Another GBV victim-survivor in Barguna, Bangladesh filed for divorce against her husband and resumed her studies. Some of the victim-survivors went to a local NGO for counselling and received help to start new lives. However, the number of women who continue to live with domestic violence is very high in Barguna society, and this number increases after each disaster.

Women also employ strategies to keep them safe from other forms of GBV, such as rape and sexual harassment. One story in particular highlights the efforts some are forced to go to. A young woman was living with her father in a house located at the bazaar. Her father had a restaurant there, and he needed help from his daughter to run the restaurant. However, she was unsafe at night due to the poor state of the housing they lived in. Local men targeted her and tried to harass her. So, she was advised by her mother to keep a knife under her bed so that she could defend herself:

> *I kept a big knife under my bed…all the time when I lived in the bazaar with my father. We had a shop in the bazaar. I helped my father. My mother said that a woman should always be prepared for any attack when she lives in the bazaar.*
>
> *(Female, Barguna)*

This woman was prepared for the worst physical attack. However, educated working women who live in better housing also commonly take preventive measures against verbal abuse and harassment. One of our key informants was an unmarried, middle-aged woman who had chosen to be single, which is very uncommon in Barguna society. Her choice always led her to be the target of social judgement, and created much curiosity. So she tried to hide it when visiting communities in rural areas as part of her work. She wears a big bindi (a small red spot) on her forehead, which is mostly worn by married Hindu women and follows the Hindu religion. This way, people do not talk about her marital status, and she can avoid the social gender-based harassment that single women receive:

> *My job is at the field level and I have to work with local inhabitants. I am unmarried. It's a common query from my respondents which I have to face every day – am I married or not? In our society a married women can avoid extra curiosity and other gender-based harassment. So, I wear a big tip (bindi) to avoid all queries. It shows I am married.*
>
> *(Female, Barguna)*

Women always have to be prepared to keep themselves safe from GBV in remote areas in Barguna and Rangpur. It is seen as their responsibility to be modest in the face of domestic violence, to cope with abusive husbands, to be prepared to fight back during physical attacks, to take shelter, to bow to cultural attitudes and above all, when facing the mighty cyclones, to be mindful of their gender identity, roles and responsibilities.

Reporting gender-based violence to the authorities

As we reflected in Chapter 4, all forms of GBV are seriously under-reported in the UK. For women in remote rural regions of Bangladesh, reporting GBV is a more complex and fraught matter still. Some women complain about their husbands' domestic abuse to local government officials, elected local Members or Chairmen when the violence becomes intolerable. Here, it should be remembered that in Bangladeshi villages, houses are located very close together and their inhabitants maintain close interdependent relationships. Families know each other, and as Members or Chairmen are elected by the villagers, they are also very well known to them; complaining to them is like complaining to a guardian of the village. When victim-survivors complain about domestic violence to the local Member or Chairman, they talk with both husband and wife and their priority is to try to keep the relationship going. They advise and sometimes threaten the man that he should manage his anger or else law-enforcement agencies will be informed. As we have discussed in previous chapters, however, most husbands do not listen to them, but instead, use their wives' complaints as justification for further violence. Meanwhile, the advice women receive is often highly gendered; women are told to be more tolerant and patient. One of the respondents complained to a Member about dowry violence by her mother-in-law, and was told '*Your mother-in-law has the right to talk badly of you, she can teach you, scold you but you cannot reply. You have to be patient*' (Female, Barguna). This woman did not report the violence to a higher authority. This type of advice is very common.

Many respondents mentioned that they were advised to stay at their parents' house, or simply to cope with the incidents, as '*the man will not change to become a good person*'; or that women should behave in a more mature, compromising and responsible manner. Again, society views female victim as being more likely to be wrong than her husband's family. This point was mentioned by one domestic abuse victim-survivor: '*Daughters-in-law always have more faults than parents-in-law. So, if you tell anyone they may accuse you. So what should I say?*' (Female, Barguna).

Reporting violence to the local authority, then, might have a small result but mostly is not effective and does not lead to positive outcomes for the women, and sometimes it makes the situation worse. Sometimes, very few victims report violence to a local NGO, to the police or to the Mahila Parisad (the Government Office which deals with problems related to GBV, especially wife beating and torture). According to one of the Mahila Parisad officials (a key informant of this study), they do receive reports from GBV victims, but the number is much lower than the actual number they know are out there:

> *Victims of wife battering are very common for us to see. We help them to receive treatment. There are huge number of wife beating cases in Barguna. All of these women cannot reach us, very few of them leave their husbands.*
>
> *(Female, Barguna)*

Two of the respondents also received help from the local NGO and the Mahila Parisad to get divorces from their abusive husbands, and then to continue their studies and find work.

Likewise, very few respondents reported violence to the police. Those who did so did not receive a good response but say that they were faced with demands for extortion: '*Corruption lies within the police department. They want money for any case*' (Female, Barguna). It goes without saying that poorer women do not have the money to bribe the police. They also do not have money to continue the case. '*I needed 1000 taka (10 GBP) to continue my case but I did not have that amount of money. So I could not do anything to get justice*' (Female, Barguna). Respondents mentioned that they would have liked to file a case against their husband and his family; however, due to lack of finances, they cannot think of taking legal action.

It is also the case that some women do not know where to report GBV, or how to get justice. They might know the name of the institutions, but this information is not enough as they don't know the steps they must take. Adding this to financial constraints and social disapproval, the context around many victims makes them feel that there is little they can do but continue to tolerate or cope with violence, making only small acts of resistance within the home.

7.3 The limits of formal responses to layered disasters

While victim-survivors are the first responders to GBV after disasters, usually in highly constrained and unsympathetic contexts as we have discussed above, we now turn to formal responses to the problem. Here, we briefly summarise feminist and related research on GBV and disasters which critically assesses responses by states, local governments and NGOs and highlights the limits of current policy and practice.

It is clear that the highly gendered impacts of disasters on women and girls (and, sometimes, men and boys) constitute one of the most significant aspects of any disaster. Referring to this as a 'double disaster' for those affected, Bradshaw and Fordham (2015) describe it as paradoxical that gender is still commonly excluded from disaster responses and risk reduction strategies. They argue that all policies around disaster planning and management should instead be 'engendered', similar to the argument often made for development and other forms of planning. In other words, gendered impacts such as increased GBV should become central, rather than peripheral, to formal disaster responses. Women and girls must not be treated as special cases in disaster management who require separate consideration, but must be the main focus of effective responses.

As Yadav et al. (2021) note, gender is recognised in risk reduction strategies to some extent, but never in ways that tackle the roots of structural inequality in disasters. They give the example of current policies in Nepal that work to protect the 'gender status quo' in the region. Instead, they argue that a feminist vision for transformative change is required that protects women, girls and sexual minorities. Similarly, George (2014) shows in her study of women's security from violence in the Pacific Islands that while 'hot' violence, or immediate crises, always attract attention, less visible forms of 'slow violence' such as sea-level rise, masculinised politics and militarism are often ignored as causal in women's safety. Likewise, as our research in Bangladesh and elsewhere shows, there is a failure to properly see and act upon the scale and effects of GBV in preparations for and responses to cyclone disasters, or the ways that GBV is closely related to climate change and ecological harm (Rezwana and Pain 2020).

Examining the case of sexual violence during disasters, Seager (2014) notes the well-documented rises during and following disasters and in humanitarian settings, and the devastating consequences for women and girls including injuries, pregnancy, HIV and other infections. She also observes that sexual violence, in creating widespread anxiety and fear, is a major factor impeding local recovery (in which women often play a central role). Seager analyses the history of demands for gender-aware policy, noting that the 2005 Hyogo Framework required the integration of gender equity into all decision-making and planning processes around disaster risk management. She reports that the actual integration of gender equity issues since then has been 'dismal', giving wide-ranging examples. As Sultana (2021) demonstrates, it is not only gender equity that must be made central to disaster planning. An intersectional approach that also focuses on race, ethnicity, class, disability, age, sexuality, religion, caste, livelihood and migrant status is essential, as all of these intersect to shape the inequalities that are compounded by disasters. Where these are not taken into account, policies will further marginalise people and reproduce existing inequalities, as Sultana (2021) demonstrates in relation to formal responses to the Covid-19 pandemic.

Around the world, from its onset in 2020, the Covid-19 pandemic led to greater awareness and discussion of the importance of centring gender in disaster responses, because of its highly gendered impacts including increases in GBV in every country (see Chapter 5). Javed and Chattu (2020) quote United Nations Chief Antonio Guterres, who stated that Covid-19 exposed what we already knew, that a male-dominated culture around the world is damaging to everyone. It is only to be hoped that this awareness does not dissipate as the pandemic recedes, but there are already grounds for doubting how far it will be sustained. While the needs of GBV victim-survivors were mostly ignored in recovery plans from Ebola, a previous recent pandemic, Javed and Chattu (2020) argue that they must be centre stage in the recovery from Covid-19. For Women's Policy Group Northern Ireland (2020), too, gender-sensitive recovery from the pandemic is essential. However, Swaminathan (2021), drawing on UN data, suggests that only 52 countries have integrated measures to prevent GBV in their Covid-19 response plans; and some have actually diverted resources from tackling GBV to deal with Covid-19. She notes that the impacts of increased GBV on mental health and on growing social and economic inequalities connected to the pandemic are rarely acknowledged or dealt with either. The long-term harm of GBV during the pandemic will be massive (Swaminathan 2021), and this relationship is cyclical, as increased poverty and unemployment and reduced access to support services are all likely to increase rates of GBV in future. This provides a compelling example of the ways that GBV both arises from and compounds crises such as Covid-19 so that they become layered disasters.

7.4 Recommendations for policy and practice

In light of the research findings we have presented in this book, and building on the material presented so far in this chapter, this section contains recommendations for disaster planning, response and recovery. Breaking the link between GBV and disasters is challenging; indeed, reducing or eradicating GBV generally is complex when the societies we live in continue to be patriarchal and riven with inequalities, not just between women and men but also between rich and poor, ethnic majority and minority groups, and so on. Gender discrimination is longstanding and embedded in all cultures and societies; gendered attitudes have become so naturalised (Bolin et al. 1998; Bourdieu 1977) for our research participants that when our research studies asked for opinions on how GBV victim-survivors might be helped during disasters, relatively few respondents were able to answer. We include their victim-survivor views in the recommendations below, but also draw on our wider research findings; GBV increases in disasters because of poverty, dependency on relief and help from others, poor quality and damaged housing, unequal power relations between family members, a lack of female empowerment, the absence of parents from the home or shelter for long hours during the day and powerful neighbours.

The recommendations are as follows.

Poverty alleviation

As this book shows that the increase of domestic violence in disasters (cyclones, floods and Covid-19) is compounded (although not solely caused) by poverty, ensuring income (for both men and women) during and after disasters will have a positive impact on GBV conditions (UNWOMEN 2014). Poverty alleviation measures should be taken to reduce women's vulnerability to both disasters and GBV. Assurance of adequate supply of food, healthcare and shelter will decrease people's struggles during periods of disasters, as well as decrease conflicts between family members. In Bangladesh, it is expected that dowry violence against women, child marriage for young girls and the migration of children will be reduced in numbers when family incomes are assured during flooding or cyclone events. As croplands remain inundated for months and crops are damaged, incentives and loans for farmers would help them to recover from the loss. Fishing equipment can also be offered to men to help them earn temporarily.

In the UK, it is clear that poverty might not cause domestic violence, but it makes it much harder for women and children to escape it. According to Lien and Lorentzen 2019, 'low income and poor educational attainments are, in themselves, risk factors for becoming the victim of violence' (Lien and Lorentzen 2019: 64). Assuring a woman's income during disasters might not only protect her from domestic violence, but help her provide for her family in the worst situations created by disasters and poverty. It offers the best chance to create resilience in the face of the consequences of domestic violence, such as wife abandonment and divorce. However, creating employment opportunities for women needs local contextual consideration (i.e. cultural, religious and social norms), especially with regard to non-conventional employment opportunities.

Hoque and Kamruzzaman (2016) found in their study that female migration may increase employment opportunities and household income, but does not necessarily change women's position either in the family or in wider society, which would be central to reducing the risks of GBV. As they observe in the case of Bangladeshi society, 'if the migrant women becomes/remains unable to take decisions or negotiate, then the "access to resources" may not be enough to reduce their vulnerabilities both in origin and destination'. They continue, 'if any policy approach, [that] claims to be gender responsive, overlooks the wider patriarchal process and structure', …'their employment in a religious mixed patriarchal structure very often interacts with particular norms and values, which act against gender equality and reinforce inequality' (Hoque and Kamruzzaman 2016: 24). So policy approaches should be gender-responsive, as well as carefully considering wider patriarchal processes and structure while operationalising any plan or programmes to involve women in income generation. In this regard, our research in disaster-prone regions of Bangladesh reveals that indoor income opportunities like sewing and making handicrafts can be arranged to assure economic stability (Chanthy and Samchan 2014). Making loans easily available from local banks to start small businesses, and introducing more vocational training for women locally, can also improve their economic condition and self-confidence. However, education can enhance women's participation in other employment sectors. Offering grants for education increases girls' presence at school. According to research on 5,000 women from very poor economic backgrounds in Bangladesh, educational programmes provide a good long-term solution that reduces domestic violence against women along with providing financial independence (Schreiber 2018). This strategy might be extended to disaster-prone regions. Besides this, reducing the number of school drop-outs among girls also has strong impacts on women's education and their participation in different types of employment sectors.

Increasing law enforcement and legal support

Increasing law enforcement units and legal support are a necessary part of the picture to protect women and girls from increases in GBV during post-disaster periods (CCJH 2018; Dutta 2017). Legal mechanisms should assist in effective disaster planning, in ways that are coordinated among cross-sectoral, national and international committees, with sufficient budgets and trained staff to ensure the availability of GBV protection after disasters (IFRCRCS 2017).

According to the present study, the lack of legal support facilities and information and the high costs of seeking this support create a barrier for poorer women who experience GBV (such as those who are divorced, victims of wife abandonment or sexual abuse) and their families. In this regard, the establishment of specialist domestic violence courts (SDVCs) and independent domestic violence advisors (IDAAs) is needed to ensure support to those at highest risk (Payton et al. 2017).

Low cost or free legal support and easily available information on these services are very helpful for these women seeking justice. Local NGOs can take these facilities and counselling support to the doorsteps of victims, in light of women's difficulties in accessing support centres. Again, protection measures and legal support should be arranged in temporary shelters to ensure the safety of women and girls during disasters as well as in post-disaster periods (DWAG 2020).

However, legal solutions do not always work for every type of GBV, as so many cases are kept hidden within the family. In particular, ending child marriage requires intervention from local government and law enforcement agencies. There are many examples of positive steps that have been taken in Bangladesh. However, there are many more cases which are not reported, and parents have even been found to fake birth certificates to marry off their daughters (see Reliefweb 2017 for more examples). Therefore, arranging grants for survivors of child marriage, laying on awareness programmes and advertising the legal support available on national TV, in local newspapers and on social media should be pursued.

Information and financial support must be provided to GBV victims, and proper implementation of laws and use of the justice system should also be encouraged. It is expected that in the long run, visible justice for victim-survivors and punishment of perpetrators will reduce rates of GBV. More examples of justice being done should lead to a change in attitudes and behaviour. Besides, awareness programmes on GBV, the laws applying to it and the legal rights of victims would raise understanding and consciousness among residents, as well as peace committees, disaster responders and other actors (First et al. 2017; USAID 2020). Assuring women's physical presence in the rural justice system also improves the situation, one example being women's presence in community justice in Rangpur (Valters and Jahan 2016).

Introducing and enhancing awareness programmes on gender and gender-based violence

Introducing and enhancing awareness programmes on gender, gender discrimination and women's empowerment help to reduce GBV (First et al. 2017; Hazeleger 2013). These programmes could cover areas such as the effects of gender discrimination, early marriage and the problems that divorced women face in conservative societies. Young girls and boys, men and women, parents and senior citizens might all be participants of such programmes. Especially, parents should be aware that discrimination between sons and daughters has a long-term effect on power relations and dynamics within their future families, and it is also important to increase awareness among parents to help daughters who experience domestic abuse and divorce. This would reduce women's vulnerability to further violence and lessen insecurity. There may also be a deterrent effect on abusive husbands, where there is less imbalance in power relations among family members. Such training should include the particular case of GBV during disasters so that families and communities are primed to look out for violence at these times, and more likely to intervene.

Involving teachers in schools and colleges in training and awareness programmes is another positive initiative to prevent violence against girls in disaster-prone regions. For example, a project in Malawi called 'Theatre for change' improved teachers' ability to protect primary school children from SGBV in flood-affected areas of Southern Malawi. This project has already achieved positive results despite some challenges and limitations (Theatre for Change 2020).

Involving local authorities in training and dialogue, such as Union Parishad Members in Bangladesh, would be very helpful to disseminate awareness on GBV and protect women and girls during disasters. For example, Concern Worldwide and ASOD (Assistance for Social Organization and Development) organised a series of dialogues between the Union Parishad (Council) Standing Committee for Women and Child Welfare and the CRAGs among 22 flood-affected communities in Bangladesh (Naznin et al. 2021). This opportunity helped to create a platform for the local authorities to share their experiences, and in making context-based recommendations to improve existing programmes as well as reducing rates of GBV in the locality. Training to service providers (counsellors, healthcare providers and professionals working in law and enforcement) is also important (Dawson et al. 2017a) because sometimes, women are harassed by NGO and government officials while receiving relief or assistance after disasters (as described in Chapters 2 and 3). Asad and his team also revealed similar incidents of violence against women while studying floods in Northern Bangladesh (Azad et al. 2013).

Men as allies

Ultimately, in order to challenge stereotypes and unequal gender roles and overcome patriarchal cultures, it is necessary to engage men and boys as allies (Javed and Chattu 2020). Men becoming women's allies, and participating in movements, training and programmes, has always been a challenge in many societies. However, from the rise of the 1970s women's anti-violence movement, the first groups of male allies formed and attended the first Take Back the Night march in the US in 1978. Messner et al. (2015) describe the experiences of three generational cohorts of men in the 1970s, 1980s and 1990s and the challenges they face in working against sexist violence (Messner et al. 2015). However challenging, men's active participation is necessary to stop violence against women. Kilmartin explains, 'in the quest to reduce men's violence, one of the greatest and perhaps most underutilised forces is the amplification of the voices and efforts of the vast majority of normal and healthy men to use their influence positively' (Kilmartin 2015: 15).

Our research in Bangladesh revealed eagerness among some male respondents to protect women from GBV. One of the male respondents in Barguna mentioned that during a recent cyclone, three men in the shelter had attacked a young girl, trying to molest her. A group of other men protested and beat the three men away. Another male respondent was found to be very sensitive and sympathetic. He also saw a man at the cyclone shelter molesting a woman, and stopped and threatened the man. He did not tell anyone about the incident, in case the woman lost her honour and was thrown out by her husband.

However, men need the support of initiatives to break down stigma and change social attitudes and behaviours around GBV. High numbers of male involvement in awareness training on disasters, GBV and their impacts on women and society are needed. The content must be mindful of Bishop's (2002) warning that we should avoid a 'hit and run pop culture approach', and rather focus on a 'structural and historical approach' to understand oppression. She warns that if we do not work from the primary principle of educating allies and teach them to 'unlearn' the oppressor's roles they have picked up from their social environment, then 'doing nothing is sometimes better' (Bishop 2002: 127). According to Michau et al. (2015), collective activism involving both women and men at all levels of the ecological model is necessary for transforming power, reducing gender inequality and preventing violence against women.

Community involvement and women's participation

In the UK, while disaster planning has for many years involved the local community level, it has seldom covered the heightened risks of GBV during disasters. Several writers on this topic recognise the power of community-based responses, groups sometimes pre-existing and sometimes forming spontaneously to tackle common problems together (Ensler 2021; Katz 2008; Seager 2014). However, informing and supporting these groups in relation to GBV needs to become integrated into existing plans and responses.

As we discussed in Section 7.2, GBV victim-survivors in Barguna and Rangpur (Chapters 2 and 3) share their sorrows and sadness with other women they know and take advice and shelter in neighbours' houses to protect themselves from domestic violence and its after-effects. The Cumbria research (Chapter 4), meanwhile, evidenced numerous examples of strong community support and mutual aid during the 2015 floods (see Deeming et al. 2015). Indeed, respondents felt that this had been more important, and more effective, than any support they received from formal actors:

> It was great to see how a community pulled together very quickly to help people who were so badly affected by the floods.
>
> Although the flooding was extremely stressful, it brought together neighbours and communities – people were so supportive and wanting to help.
>
> We're Cumbrians… we are a working community and everyone just got on with the recovery.
>
> *(Females, Cumbria)*

However, for those few who reported it to the research, GBV was rarely talked about within the community, and they felt alone in this aspect of the disaster. Disaster planners must therefore recognise the powerful capacity and knowledge of communities and involve them further in planning, but in ways that account for and address inequalities of race, class and gender in order to achieve socially just outcomes (Jacobs 2019). The awareness and engagement of local communities in this way can act to protect those who suffer most when disasters strike, including the specific risks of GBV (Ertan et al. 2020; World Health Organization 2020). The Covid-19 pandemic has recently opened more space in higher-income countries for wider community engagement in tackling GBV, for example, in bystander intervention and raised awareness and training in sites outside the sphere of criminal justice (Walklate et al. 2020); this might be directly built into disaster planning.

There are many calls from feminists, too, for women to have opportunities to fully participate in developing disaster plans and responses. Ideally, women, and especially minoritised women, should hold positions of leadership and inform policies to respond to GBV in disasters (Drew 2020). As Bradshaw and Fordham (2015) point out, women may have many ideas and skills when it comes to adaptation and mitigation of disasters; but they generally have less power than men within communities and are more time-poor and constrained. It should also be kept in mind that while it is ideal that women participate equally in disaster recovery on disaster committees, receiving loans, attending training and community recovery programmes, inclusion in recovery programmes should not disproportionately fall on women, increasing their responsibilities and the challenges they face during the post-disaster period (Enarson 2014). Therefore women's participation, in particular, needs to be properly supported with adequate compensation, training and mentoring.

A positive aspect of this involvement is that, in itself, it helps to protect women from some of the harmful impacts of disasters; research shows that gaining social capital also helps women to gain empowerment and self-confidence (Ganapati 2012).

Healthcare facilities for GBV victim-survivors in disasters

Good healthcare facilities and access, both for physical and psychological health, are essential for GBV victim-survivors to continue living healthy lives. The studies reported in Chapters 2 and 3 revealed that many live with physical injuries and mental health impacts from GBV due to a lack of healthcare during disasters. For example, during floods, movement becomes restricted due to inundated roads and a lack of transport such as boats and rafts. Response programmes need to arrange boats to facilitate movement, especially taking patients to healthcare centres.

Social attitudes towards GBV victim-survivors also suggest the need for separate sections in healthcare centres or separate facilities. For example, OCCs (One-stop Crisis Centres) have been established in Bangladesh since 2000: there are now six OCCs (MSPVAW 2022) which are very helpful to female victim-survivors of GBV. However, they are insufficient in numbers and many more centres like this are needed at the local level, especially during disasters.

According to Michau et al. (2015), the health sector should be strengthened and its role in preventing violence against women increased. Healthcare providers are the first point of contact for GBV victims and they should be trained to identify and respond to these emergencies. Health facilities also have an important role in referrals to specialist abuse services (World Health Organization 2020). It is important that both physical and mental health conditions are prioritised in these plans (USAID 2020), as women at most risk of disasters find psychological support helpful during the difficult post-disaster period (Shooshtari et al. 2017). Ganapati (2012) shows how public administration helped to build social capital in this way, to support women recovering after disasters in Turkey.

In another example, United Nations Population Fund (UNFPA) provides safe spaces, mobile clinics and mobile outreach teams to bring psychological counselling and life-saving services and supplies to adolescent girls who are displaced by conflicts or disasters in Malawi, Nepal, Nigeria, Pakistan, Myanmar and Somalia (UNFPA 2016). These interventions were proven to have positive impacts on girls in humanitarian settings. Shelter Cluster Bangladesh (2018) recommends that the contact details of formal and informal psychological support systems should be available to make referrals when necessary. Fisher (2010) also suggests that men's health and psychological impacts of disasters on men, such as stress and trauma, alcohol consumption and drug abuse, should be addressed by disaster response plans.

Gender-sensitive disaster management plans to improve GBV prevention and response

Mainstreaming a gender perspective and introducing gender-sensitive disaster management plans for pre-, during and post-disaster periods is essential to reduce the vulnerability of women and girls to disasters and GBV (DWAG 2020; Fisher 2010; IFRCRCS 2017; Rezwana 2018; UNWOMEN 2014). According to the Council of Europe, gender mainstreaming is '(re)organization, improvement, development and evaluation of the political processes, so that the gender equality perspective is incorporated into all policies, at all levels and in all the stages by the actors normally involved in making political decisions' (Council of Europe 1998: 15 in Akerkar and Fordham 2017: 227). The research reported in this book reveals increases in GBV during disasters; GBV takes place as women and girls go on their way to take shelter, at cyclone shelters and at their damaged houses during the post-disaster period. Women's security must be tackled during these times by training state and non-state actors. Far wider recognition is needed that GBV often increases in numbers and severity during and after disasters (CCJH 2018; UNWOMEN 2014), and formal responders should have clear knowledge on GBV types, forms and contexts. At present, GBV often does not get proper attention. According to Parkinson and Zara (2013), emergency management in Australia during the recovery phase of the 2009 bushfires included counselling services for individual psycho-social needs such as grief and loss, but domestic violence was not recognised as a relevant issue.

Gender is, however, mentioned as a cross-cutting principle for building disaster resilience in the Hyogo Framework for Action 2000–2015 (Dutta 2017). We have already mentioned Seager's (2014) appraisal of the actual application of this principle as 'dismal'. According to Enarson (2014), far more is needed: a gendered human security framework should be introduced in disaster planning which allows 'women and men equally to anticipate, resist, survive, cope with, and recover from the effects of hazards and disasters' (Enarson 2014: 50). Michau et al. (2015) recommend that policy makers, funders, service providers, and programmers should follow principles of effective prevention of violence against women and girls. The principles should be based on 'years of practice' and 'collaborative practitioner-researcher impact evaluations', to ensure 'future well targeted investments and meaningful knowledge exchange' to prevent violence against women (Michau et al. 2015: 1681). According to Yadav et al. (2021), applying a feminist vision is needed to recognise biases and problems in prevailing DRR policies and practices. Again, to understand the context of social norms and culture in relation to GBV in disasters and its material costs (UNICEF 2017), more cross-sectoral innovations for plans and policies are necessary, as are investment and funding opportunities in innovation to enhance gender sensitivity in disaster response plans (Michau et al. 2015). According to a UNICEF Report (2017), sovereign debt, blended finance, climate finance and project finance investments all have the potential to help in addressing or reducing vulnerability to GBV.

However, it is also recommended that local cultures and contexts of GBV should be prioritised (Buscher 2014; Masson et al. 2016), and the concerns of local actors and 'the limited capacities of international organizations' should be considered in preparing gender-sensitive disaster management plans. Of course, 'adapting universal principles to local conditions and translating them into local idioms' (Abramowitz and Moran 2012: 140) is not always straightforward, but it is very important (Dawson et al. 2017b), because the 'cultural context can contribute to both a woman's risk of GBV and resiliency in surviving GBV' (O'Brien and Macy 2016: 49). For example, the UNFPA implemented Adolescent Sexual and Reproductive Health in Emergencies (ASRHiE) strategy in the Philippines involved adolescents and youth, especially females, in the humanitarian response as first responders and volunteers. They were consulted and engaged in planning, distributing relief and communicating with local communities, and 'young people proved to be valuable partners as first responders and outreach volunteers in the humanitarian response' in the Philippines (UNFPA 2016: 74).

In another example, Shelter Cluster Bangladesh (2018) recommends involving women and women's organisations in the sitting, design, construction, management and coordination of shelter facilities in Bangladesh. Fisher (2010), in her paper on post-tsunami conditions in Sri Lanka, recommends the involvement of experienced local providers and women's groups in response plans, as they have the capacity to respond appropriately to culturally sensitive issues like GBV. It was noted in Chapter 3 that some local women in Rangpur actively work to increase awareness of GBV among the local population. Issues such as domestic violence are not just something to be dealt with by law and enforcement, healthcare providers or GBV specialists, but 'they are a challenge for us all, as a society, to work on in productive collaboration' (Payton et al. 2017: 116). The most crucial contribution of culture in preventing violence against women is enhancing a collaborative ethos among members of society (Payton et al. 2017), which is helpful in implementing gender-sensitive disaster management plans and programmes effectively. Some examples of disaster-sensitive planning that are relevant to the research reported in this book are as follows.

Local volunteers, who are a great help during cyclones (Rezwana 2018) in disseminating warnings and rescuing and taking vulnerable people to the safe shelters, can receive training to protect women and girls who take shelter in the cyclone shelters. Here, gender-disaggregated information will be helpful for the responders (UNWOMEN 2014; USAID 2020). For example, the Victorian Action Plan to Address Violence Against Women and Children includes emergency management workers in Common Risk Assessment for Family Violence training (Hazeleger 2013). Awareness training is arranged for the volunteers and cyclone shelter management committees. Without such training, frontline volunteers may face unexpected situations, leading to confusion and lack of self-confidence and negative impacts on their psychological health (Deeming et al. 2015).

Secondly, flood shelters should be safe for women. There are currently no security systems in the shelters commonly used by respondents in Bangladesh. Arranging separate entry and exits, rooms and permanent and temporary toilets (USAID 2020) for males and females in the disaster shelters, and proper lighting at night, will decrease the incidence of GBV (Shelter Cluster Bangladesh 2018). In particular, more shelters are needed if further floods or cyclones hit. The dissemination of clear information before disasters on their availability will reduce confusion. More shelters will assure more space, which is also helpful to combat Covid-19 and other infectious diseases, as well as GBV. For example, during Cyclone Amphan, which hit Bangladesh in 2020, according to the CPP (Cyclone Preparedness Programme) report, coordination between the Ministry of Disaster Management and Relief (MoDMR), local government, the district administration, Ansar/Village Defence Party (VDP), Cyclone Preparedness Programme(CPP) and other humanitarian actors worked well during Cyclone Amphan to arrange more safe shelters (CPP 2021).

In other disasters, it has become clear that women and girls need places to go to for help that differ from the places they may go in non-disaster times. During the initial waves of Covid-19, Pentaraki and Speake (2020) report some successes in European countries using places such as pharmacies and supermarkets that remained open during the pandemic, with campaigns highlighting 'safe words' that could be used to communicate the form of help needed. In the UK, those experiencing domestic abuse were able to make silent calls to the general emergency number, tapping the receiver or pressing additional buttons to receive assistance. However, a lot of support for GBV survivors went online during pandemic, which is not always suitable for them, especially as perpetrators of domestic abuse may also be in the house and commonly monitor digital activity closely. In addition, lower-income groups may not have suitable internet or phone access (Pentaraki and Speake 2020). In future pandemics and other disasters, specialist domestic and sexual violence support services need to be considered as essential services and should be made fully accessible with their physical premises staying open (World Health Organization 2020).

Other strategies suggested by respondents in the Bangladesh studies include measures to tackle GBV during relief distribution, which can be aided by a stronger management system. Relief for all, with an adequate amount and distribution, will decrease the likelihood of women and girls being cheated and abused. Further, as mentioned above, providing jobs for female disaster victims will enable their economic security which saves them from being trafficked for jobs or ransom marriages. Besides this, local residents of Kashiganj village also mentioned permanent solutions such as building embankments, river dredging and making efficient disaster response plans. Respondents in Cumbria in the UK also mentioned dissatisfaction with the rate at which flood defences were being built to prepare their towns for future incidents.

Gaps in knowledge

As we have discussed, a lack of data on GBV during disasters is a significant gap that affects disaster planning, but it is also needed during disasters to inform response and recovery strategies. At present, there is a serious lack of information and, even more worryingly, reports of incidents of sexual violence tend to be disbelieved (Seager 2014). As Klein (2008: 12) puts it, 'the reporting of sexual violence in disasters is often considered a luxury issue...further down on the hierarchy of needs'.

Therefore a key area of action must be prioritising gender-disaggregated data and other vulnerability indicators, in order to identify the needs of vulnerable groups (Fisher 2010). Other gaps in knowledge include how rates of violence change; the availability of disaggregated data for different groups of women that would enable an intersectional approach to responses; pathways to violence; and good practice in prevention during outbreaks (Fraser 2020). GBV data might be collected in specialist, contextualised research during disasters (Hossain and McAlpine 2017; Montesanti 2015) such as the methods used in the research studies we have reported in this book. Generally, more research at local levels is required, in order to understand the specific and contextualised nature of issues of GBV in disasters.

7.5 Conclusion: responses to layered disasters

We, like many other GBV researchers, have suggested in this chapter that it is vital to reshape formal responses to disasters so that GBV is a central consideration. Gender oppression and the urgent risks that GBV poses require intervention and prioritisation, both in disaster planning and management and in non-disaster times. At present, most interventions are inadequate to protect women, and sometimes actually worsen the impacts of disasters on them, continuing the destructive cycle of disasters, poverty and GBV. Improvements to policy and practice must also be premised on an understanding of the structural causes of GBV and must acknowledge and work with the strategies of resistance that victim-survivors already employ.

In Chapter 1, we introduced Luft's (2016) concept of racialised disaster patriarchy. Arguing that 'gendered patterns help to produce disasters and part of what is disastrous is patriarchy' (Luft 2016: 8), Luft analysed how patriarchy and racism are reinscribed through the process of disaster response and recovery. While her example is responses to Hurricane Katrina in the US, her framing has much salience with what we have described in our various case studies in this book. Luft's analysis follows through from the causes of the disaster, to immediate responses, to recovery and the rebuilding of institutions and the social contract after disasters. She shows how, in the reorganisation and redevelopment of housing, healthcare and education in New Orleans, previous laws and systems were overhauled in a move from public to private provision. This had gendered effects that disproportionately affected women.

Luft also charted resistance to these macro-level changes among community-based groups in the city, but found that these, too, had a highly gendered and intersectional nature. Hierarchies of labour within social movements emerged: men having 'big ideas' while women did much of the work on the ground. The gendering of these practices, activities and effects within social movements, Luft argues, further retrenched 'disaster patriarchy' in post-disaster formations.

She concludes that

> We cannot wait for crisis to mobilize feminist intersectional justice efforts. Instead, the groundwork must be laid before disaster strikes. For it is here, as the people of New Orleans know well, that the conditions of disaster are made.
>
> *(Luft 2016: 21)*

Chapter 5 returned to this framing of racialised disaster patriarchy as it applies to responses to the Covid-19 pandemic (Ensler 2021). Covid-19 has disproportionately affected women and girls and led to increased rates of various forms of GBV around the world. Here, we see that economic and social governance before and during the pandemic created greater risks of GBV and greater insecurity for those affected by it (Ensler 2021). The common lack of planning for GBV in any disaster planning and management was made worse during Covid-19, where lockdowns were a key policy response to the disaster (Drew 2020); the pandemic became a layered disaster which was foreseeable, but to which there was little response until it was too late for many of the women and children harmed or killed in lockdowns and the long tail of consequences that will follow. However, some of the damage was done long before the first cases of Covid-19. The rolling back of state provision, and the running down of public services in many countries in the years leading up to the pandemic, created conditions including fragile healthcare, social protection and justice systems which made the impacts of the pandemic far worse than they might have been (Javed and Chattu 2020).

This lack of social care and provision long before disasters strike is fundamental to disaster impacts. As we saw clearly from the studies in Bangladesh, if GBV is not prioritised by national or local governments in non-disaster times, and services are lacking, then it is no surprise that the problem becomes amplified during cyclones and floods. Poverty and the climate crisis are adding to this layered disaster. And in Cumbria in the UK, growing income inequalities between rich and poor, problems of insecure housing tenure, and the inability of some people to insure their houses mean that some women and girls are becoming more vulnerable to both GBV and flooding. All these examples demonstrate the powerful role of political, social and economic conditions in shaping disasters and disaster outcomes (Kelman 2020; O'Keefe et al. 1976), and also that gender inequalities produced by patriarchy must be one of the key facets in this understanding of disasters.

We end this book by refocusing our attention back on victim-survivors of GBV in disasters. While state and even NGO responses are often inadequate in tackling the challenges faced, and sometimes do more harm than good, it is clear that victim-survivors of both disasters and GBV are taking action into their own hands. Around the world, the recent Covid-19 pandemic has seen the emergence of informal collective action recognised around the world (Ensler 2021). Women, migrants and minoritised groups are often centrally involved, although there is relatively little research on their gendered aspects. Many hurriedly came together as local crisis mode responses with no clear political steer, but others have been created by social movements as intentional resistance to the violences of capitalism, racism and patriarchy. Firth (2022) describes these spontaneous movements as 'disaster anarchy' during the pandemic, charting mutual aid networks, grassroots direct action, occupations and other social movement responses. Many have filled essential gaps left by the state during the crisis (Jacobs 2021), for example, in repairing or replacing infrastructure, providing food and care-giving, and promoting resilience and survival (Auerbach et al. 2021; Eaves and Al-Hindi 2020). Sultana (2021) also notes the solidarity that became stronger during the pandemic and that this mutual aid included many national and global feminist organisations. She describes these as

> Radical intersectionality approaches [that] go beyond just analyses and explanation, but in exploring opportunities for building solidarities, agency, and collectivities against systemic inequities and oppression. It offers possibilities to explore strategies of resistances and emancipatory transformations. It also allows for better policy-making and project planning for reducing climate-related and pandemic-related vulnerabilities and crises.
>
> *(Sultana 2021: 454)*

Such mutual aid is ideally placed to tackle GBV and provide immediate support to victim-survivors in subsequent disasters. This is not to absolve state and other formal responses of responsibility, as Appiah (2021: no page number) observes, 'around these small, local circles of reciprocal caring, we need to build larger, global ones. Resilience shouldn't be reserved for the rich'. In a powerful piece on the structural causes of the pandemic, Roy (2020) calls not for a return to normal after Covid-19, but insists that a new world with different forms of organisation, one that prioritises the most vulnerable people, must be our goal.

Kofman and Garfin (2020) are hopeful that the learning that has taken place globally during the pandemic will benefit victim-survivors of GBV during future disasters. They note that specialist GBV organisations have not only innovated in their practices during Covid-19, but have compiled comprehensive DV guidelines and recommendations for response and recovery which might now be implemented at local, state, and national levels. They argue that

> Long-term, the pandemic may serve as a critical inflection point for implementing planning and preparedness guidelines to protect DV victims and survivors in the face of the ongoing threat of COVID-19 and the promise of future disasters.
>
> *(Kofman and Garfin 2020: 199)*

As community- and survivor-based approaches have shown clearly, holistic disaster response and recovery approaches are now needed that seek to reduce gender inequality, not accommodate or reproduce it (Seager 2014).

References

Abramowitz, S. & Moran, M. H. (2012). "International human rights, gender-based violence, and local discourses of abuse in postconflict Liberia: a problem of 'culture'?" *African Studies Review 55*(2): 119–146.

Akerkar, S. & Fordham, M. (2017). "Gender, place and mental health recovery in disasters: addressing issues of equality and difference." *International Journal of Disaster Risk Reduction 23*: 218–230.

Anwary, A. (2015). "Construction of hegemonic masculinity: violence against wives in Bangladesh." *Women's Studies International Forum* 50:37–46.

Appiah, K. A. (2021). "A tale of two pandemics: the true cost of Covid in the global south." *The Guardian*, 23 November 2021.

Auerbach, J., Clark, J., & Muñoz, S. (2021). "Under quarantine in a city project: stories of fear, family, food, and community." In Doucet, B., van Melik, R., & Filion, P. (eds.) *Volume 1: Community and Society*: 117–128. (Global Reflections on COVID-19 and Urban Inequalities). https://doi.org/10.2307/j.ctv1t4m1nq.17, accessed on 30/05/22.

Azad, A. K., Hossain, K. M., & Nasreen, M. (2013). "Flood-induced vulnerabilities and problems encountered by women in northern Bangladesh." *International Journal of Disaster Risk Science* 4(4): 190–199.

Bishop, A. (2002). *Becoming an Ally: Breaking the Cycle of Oppression in People*. Canada, Halifax, Farnwood Publishing.

Bolin, R., Jackson, M., & Crist, A. (1998). "Gender inequality, vulnerability, and disaster: issues in theory and research." In Enarson, E. P. & Morrow, B. H. (eds.) *The Gendered Terrain of Disaster: Through Women's Eyes*. New York, Praeger: 27–44.

Bourdieu, P. (1977). *Outline of a Theory of Practice*. Cambridge, Cambridge University Press.

Bradshaw, S. & Fordham, M. (2015). "Double disaster: disaster through a gender lens." In Collins, A. E., Jones, S., Manyena, B., & Jayawickrama, J. (eds.) *Hazards, Risks and Disasters in Society*. Cambridge, Academic Press: 233–251.

Buscher, D. (2014). "Preventing gender-based violence: getting it right." *Gender-based Violence in Emergencies*, Humanitarian Practice Network, ODI, Issue 16, Article 5, https://odihpn.org/publication/preventing-gender-based-violence-getting-it-right/, accessed on 05/02/2022.

CCJH (2018). "Gender-based violence in disasters Information Sheet." Centre for Criminal Justice and Human Rights, School of Law, University College Cork, Ireland. www.ucc.ie/en/ccjhr, accessed on 19/07/2021.

Chanthy, S. & Samchan, H. (2014). "Flood impacts on women: exploring the possibility of gender sensitive DRR planning." Actionaid, Cambodia. https://reliefweb.int/sites/reliefweb.int/files/resources/KHM_0601_ImpactonWomen_AA2014-01.pdf, accessed on 18/07/2021.

Council of Europe (1998). *Gender mainstreaming: conceptual framework, methodology and presentation of good practices*, Council of Europe, EG-S-MS (98) 2 rev.

CPP (2021). "Combating Cyclone in COVID-19 Environment: Modified Cyclone Preparedness & Early Action Plan." Cyclone Preparedness Programme (CPP), Ministry of Disaster Management and Relief, Bangladesh, www.cpp.gov.bd.

Dawson, M., Jaffe, P., Campbell, W. L., & Kerr, K. (2017a). "Canada." In Dawson, M. (ed.) *Domestic Homicide and Death Review: An International Perspective*. London, UK, Palgrave Macmillan.

Dawson, M., Mathews, S., Abrahams, N., & Campbell, J. (2017b). "Death reviews in the context of domestic homicide in low-to middle-income countries: South Africa as a case study." In Dawson, M. (eds.) *Domestic Homicide and Death Review: An International Perspective*. London, UK, Palgrave Macmillan.

Deeming, H., Fordham. M., & Davis, B. (2015). "Propagating the aftermath: the Cumbrian experience of using a community development approach to build resilience to flooding". Paper presented at: *Social Services in Times of Disaster, NORDRESS Multidisciplinary Nordic Symposium*, Nordic House, Reykjavik, Iceland.

Drew, M. (2020). "Feminist perspectives on disaster, pandemics, and IPV." In Geffner, R., White, J. W., Hamberger, L. K., Rosenbaum, A., Vaughan-Eden, V., & Vieth, V. I. (eds.) *Handbook of Interpersonal Violence across the Lifespan*. New York, Springer: 1–20.

Dutta, P. M. (2017). "Natural disaster and vulnerability to trafficking of women and girls in India." *European Scientific Journal 7881*: 54–69.

DWAG (2020). "Ongoing floods expose women to further risk of gender-based violence." http://www.darfurwomenaction.org/press_release/ongoing-floods-expose-women-to-further-risk-of-gender-based-violence/, accessed on 06/06/2021.

Eaves, L. & Al-Hindi, K. F. (2020). "Intersectional geographies and COVID-19." *Dialogues in Human Geography* 10(2): 132–136.

Enadner, V. & Holmberg, C. (2008). "Why does she leave? The leaving process(es) of battered women." *Health Care for Women International* 29: 200–226.

Enarson, E. (2014). "Human security and disasters: what a gender lens offers." In Hobson, C., Bacon, P., & Cameron, R. *Human Security and Natural Disasters*. Oxon, Routledge: 37–56.

Enarson, E. & Chakrabarti, P. D. (eds.) (2009). *Women, Gender and Disaster: Global Issues and Initiatives*. New Delhi, India, SAGE Publications.

Enarson, E., Fothergill, A., & Peek, L. (2018). "Gender and disaster: foundations and directions for research and practice." In Rodriguez, H., Donner, W., & Trainor, J. E. (eds.) *Handbook of Disaster Research*. New York, Springer: 205–223.

Ensler, E. (2021). "Disaster patriarchy: how the pandemic has unleashed a war on women." *The Guardian*, 1 June 2021.

Ertan, D., El-Hage, W., Thierrée, S., Javelot, H., & Hingray, C. (2020). "COVID-19: urgency for distancing from domestic violence." *European Journal of Psychotraumatology* 11: 1.

First, J. M., First, N. L., & Houston, J. B. (2017). "Intimate partner violence and disasters: a framework for empowering women experiencing violence in disaster settings." *Affilia* 32(3): 390–403.

Firth, R. (2022). *Disaster Anarchy: Mutual Aid and Radical Action*. London, Pluto Press.

Fisher, S. (2010). "Violence against women and natural disasters: findings from post-tsunami Sri Lanka." *Violence Against Women 16*(8): 902–918.

Fraser, E. (2020). "Impact of COVID-19 Pandemic on Violence against Women and Girls." *VAWG Helpdesk Research Report no. 284*, London, UKAid.

Ganapati, N. E. (2012). In good company: why social capital matters for women during disaster recovery. *Public Administration Review* 72(3): 419–427.

George, N. (2014). "Promoting women, peace and security in the Pacific Islands: hot conflict/slow violence." *Australian Journal of International Affairs* 68(3): 314–332.

Hazeleger, T. (2013). "Gender and disaster recovery: strategic issues and action in Australia." *The Australian Journal of Emergency Management* 28(2): 40–46.

Hoque, M. M. & Kamruzzaman, M. (2016). *From kitchen utensil to sewing machine: women's internal migration and gender dynamics in rural Bangladesh*. GIGD, BRAC University, Bangladesh.

Hossain, M. & McAlpine, A. (2017). *Gender Based Violence Research Methodologies in Humanitarian Settings: An Evidence Review and Recommendations*. Elhra, Cardiff.

IFRCRCS (2017). "Effective law and policy on gender equality and protection from sexual and gender-based violence in disasters." www.ifrc.org, accessed on 18/04/2021.

Jacobs, F. (2019). "Black feminism and radical planning: new directions for disaster planning research." *Planning Theory* 18(1): 24–39.

Jacobs, F. (2021). "Beyond social vulnerability: COVID-19 as a disaster of racial capitatlism." *Sociologica* 15(1): 55–65.

Javed, S. & Chattu, V. J. (2020). "Patriarchy at the helm of gender-based violence during COVID-19." *Aims Public Health* 8(1): 32–35.

Katz, C. (2008). "Bad elements: Katrina and the scoured landscape of social reproduction." *Gender Place and Culture* 15(1): 15–29.

Kelman, I. (2020). *Disaster by Choice: How Our Actions Turn Natural Hazards into Catastrophes*. Oxford, UK, Oxford University Press.

Kilmartin, C. (2015). "Men's violence against women: an overview." In Johnson, A. J. (eds.) *Religion and Men's Violence Against Women*. New york, Springer: 15–25.

Klein, N. (2008). *The Shock Doctrine: The Rise of Disaster Capitalism*. London, Penguin.

Kofman, Y. B. & Garfin, D. R. (2020). "Home is not always a haven: the domestic violence crisis amid the COVID-19 pandemic." *Psychological Trauma: Theory, Research, Practice, and Policy* 12(S1): S199–S201.

Lien, M. I. & Lorentzen, J. (2019). *Men's Experiences of Violence in Intimate Relationships*. Switzerland, Palgrave Macmillan.

Luft, R. E. (2016). "Racialized disaster patriarchy: an intersectional model for understanding disaster ten years after Hurricane Katrina." *Feminist Formations* 28(2): 1–26.

Masson, V. L., Lim, S. Budimir, M., & Podboj, J. S. (2016). "Disasters and violence against women and girls. Can disasters shake social norms and power relations?" London, Overseas Development Institute.

Messner, M. A., Greenberg, M. A., & Peretz, T. (2015). *Some Men: Feminist Allies in the Movement to End Violence Against Women*. New York, Oxford University Press.

Michau, L., Horn, J., Bank, A., Dutt, M., & Zimmerman, C. (2015). "Prevention of violence against women and girls: lessons from practice." *The Lancet* 385(9978): 1672–1684.

Mileti, D. (1999). *Disasters by Design: A Reassessment of Natural Hazards in the United States.* Washington DC, Henry Joseph Press.

Montesanti, S. R. (2015). "The role of structural and interpersonal violence in the lives of women: a conceptual shift in prevention of gender-based violence." *BMC Women's Health* 15(1): 1–3.

MSPVAW (2022). "One-Stop Crisis Centre (OCC)." http://mspvaw.gov.bd/contain/15, accessed on 01/06/2022.

Naznin, Z., Sharmin, M., & Mohan, F. (2021). Addressing gender-based violence in Bangladesh. Flood Resilience Portal, Concern worldwide, Bangladesh. https://floodresilience.net/blogs/addressing-gender-based-violence-in-bangladesh/, accessed on 21/09/2021.

Nasreen, M. (2008). *Violence against Women during Flood and Post-flood Situation in Bangladesh.* Bangladesh, ActionAid.

O'Keefe, P., Westgate, K., & Wisner, B. (1976). "Taking the naturalness out of natural disasters." *Nature* 260: 566–567.

O'Brien, J. E. & Macy, R. J. (2016). "Culturally specific interventions for female survivors of gender-based violence." *Aggression and Violent Behavior* 31: 48–60.

Pain, R. (2014). "Seismologies of emotion: fear and activism during domestic violence." *Social & Cultural Geography* 15(2): 127–150.

Parkinson, D. & Zara, C. (2013). "The hidden disaster: domestic violence in the aftermath of natural disaster." *The Australian Journal of Emergency Management* 28(2): 28–35.

Payton, J., Robinson, A., & Brookman, F. (2017). "United Kingdom." In Dawson, M. (ed.) *Domestic Homicide and Death Review: An International Perspective.* London, Palgrave Macmillan: 91–124.

Pentaraki, M. & Speake, J. (2020). "Domestic violence in a COVID-19 context: exploring emerging issues through a systematic analysis of the literature." *Open Journal of Social Sciences* 8: 193–211.

Reliefweb (2017). "Stopping child marriage forever." www.reliefweb.int/report/bangladesh/stopping-child-marriage-forever, accessed on 01/06/2022.

Rezwana, N. (2018) *Disasters, Gender and Healthcare Access: Women in Coastal Bangladesh.* Oxon, UK, Routledge, Taylor& Francis.

Rezwana, N. & Pain, R. (2020). "Slow violence and layered disasters: gender-based violence before, during and after cyclones." *Disasters* 45 (4): 741–761.

Roy, A. (2020). "The pandemic is a portal." *Financial Times.* https://www.ft.com/content/10d8f5e8-74eb-11ea-95fe-fcd274e920ca, accessed on 15/06/2021.

Schreiber, M.. (2018). "Can a woman's rising social status bring down rates of domestic violence?" NPR. http://www.npr.org/, accessed on 12/05/2020.

Seager, J. (2014). "Disasters are gendered: what's new?" In Singh A. & Zommers, Z. (eds.) *Reducing Disaster: Early Warning Systems for Climate Change.* Dordrecht, Springer: 265–281.

Shelter Cluster Bangladesh (2018). Gender and diversity in emergency shelter programming. https://www.humanitarianresponse.info/sites/www.humanitarianresponse.info/files/documents/files/scbang1.pdf, accessed on 11/10/2021.

Shooshtari, S., Abedi, M., Bahrami, M., & Samouei, R. (2017). "Women's mental health as the basis for preventive planning for disasters." *Journal of Safety Promotion and Injury Prevention* 5(2): 61–62.

Sultana, F. (2021). "Climate change, COVID-19, and the co-production of injustices: a feminist reading of overlapping crises." *Social & Cultural Geography* 22(4): 447–460.

Swaminathan, S. (2021). "Covid-19 is fueling a pandemic of violence against women and girls." *International Science Council Blog*. https://council.science/current/blog/covid-19-is-fueling-a-pandemic-of-violence-against-women-and-girls/, accessed 07/04/22.

Theatre for Change (2020). Improving the ability of teachers to protect primary school children from sexual and gender-based violence in flood affected areas, Year One Narrative Report for GIZ Oct 2019 – Sept 2020. https://healtheducationresources.unesco.org/sites/default/files/resources/23605_GIZ-Year-1-Narrative-Report-9.11.20.pdf, accessed on 21/09/2021.

UNFPA (2016). Adolescent Girls in Disaster & Conflict: Interventions for Improving Access to Sexual and Reproductive Health Services, United Nations Population Fund. https://www.unfpa.org/sites/default/files/pub-pdf/UNFPA-Adolescent_Girls_in_Disaster_Conflict-Web.pdf, accessed on 12/10/2021.

UNICEF (2017). The material risks of gender-based violence in emergency settings: is GBV impacting your investment? www.gender@unicef.org, accessed 07/03/2021.

UNWOMEN (2014). Climate change, disasters and gender-based violence in the pacific. www.asiapacific.unwomen.org, accessed on 20/03/2021.

USAID (2020). Feed the Future Ethiopia Resilience in Pastoral Areas (RiPA North), Rapid Gender Analysis for Afar Flood. https://pdf.usaid.gov/pdf_docs/PA00X2TX.pdf, accessed on 25/09/2021.

Valters, C. & Jahan, F. (2016). *Women and Power, Mediating community Justice in Rural Bangladesh*. London, Overseas Development Institute.

Walklate, S., Richardson, J., & Godfrey, B. (2020). *Domestic Abuse-Family Violence, Disasters and Restrictions under Covid-19: An Overview*. Working Paper No. 1. Domestic Abuse: Responding to the Shadow Pandemic, University of Liverpool, UK.

Women's Policy Group Northern Ireland (2020). Covid-19 Feminist Recovery Plan. https://wra.net/wp-content/uploads/2020/07/WPG-NI-Feminist-Recovery-Plan-2020.pdf, accessed on 07/04/22.

World Health Organization (2020). *COVID-19 and violence against women: What the health sector/system can do*. WHO, Geneva.

Yadav, P., Saville, N., Arjyal, A., Baral, S., Kostkova, P., & Fordham, M. (2021). "A feminist vision for transformative change to disaster risk reduction policies and practices." *International Journal of Disaster Risk Reduction* 54: 102026.

INDEX